D0821831

Eureka Man

BY THE SAME AUTHOR

The Electric Life of Michael Faraday
Parallax: The Race to Measure the Cosmos
Astronomy Activity and Laboratory Manual

Eureka Man

The Life and Legacy of Archimedes

Alan Hirshfeld

Walker & Company
New York

Published by Walker Publishing Company, Inc., New York

All papers used by Walker & Company are natural, recyclable products made
from wood grown in well-managed forests. The manufacturing processes
conform to the environmental regulations of the country of origin.

Every reasonable effort has been made to contact copyright holders of images
reproduced in this book, but if any have been inadvertently overlooked the
publishers would be glad to hear from them and to make good in future
editions any errors or omissions brought to their attention.

LIBRARY OF CONGRESS CATALOGING–IN–PUBLICATION DATA

Hirshfeld, Alan.
Eureka man : the life and legacy of Archimedes /
Alan Hirshfeld.
p.   cm.
Includes bibliographical references and index.
ISBN 978-0-8027-1618-7 (hardcover)
1. Archimedes—Biography. 2. Scientists—Greece—Biography. I. Title.
Q143.A62H57 2009
509.2—dc22
[B]
2009005608

Visit Walker & Company's Web site at www.walkerbooks.com

First U.S. edition 2009

1 3 5 7 9 10 8 6 4 2

Typeset by Westchester Book Group
Printed in the United States of America by Quebecor World Fairfield

To Sasha, Josh, and Gabe

CONTENTS

Part I

MASTER OF THOUGHT

Chapter 1

THE ESSENTIAL ARCHIMEDES

There was more imagination in the head of Archimedes
than there was in that of Homer.

—VOLTAIRE, *PHILOSOPHICAL DICTIONARY*

THE CITIZENS OF ancient Syracuse would have recognized the man who is said to have bustled past them naked and dripping and shouting, "*Eureka!*" (I have found it!). It was Archimedes, the celebrated mathematician, scientist, inventor, and confidant of the king. That Archimedes seemed oblivious to his own nakedness and to onlookers' bemused stares was perhaps only mildly scandalous, given his reputation for eccentricity. To residents of this long-ago Sicilian city-state, Archimedes had always occupied an enchanted middle ground: one foot planted squarely in the world of men, the other dancing to some private muse of nature.

Archimedes was naked and wet because, only moments earlier, he had purportedly jumped from his bath, elated at his flash of insight into a problem he had been puzzling over. The Syracusan king, Hieron II, had given the royal metalsmith a specific weight of gold to be fashioned into a splendid wreathlike crown. Now the king suspected that the completed crown, destined to adorn the statue of a deity, had been cut with less valuable silver and that the smith had pocketed the unused gold. Hieron directed Archimedes to establish the crown's makeup without sampling or defacing it in any way.

Archimedes knew that gold is more dense than silver. So if a certain weight of silver had been substituted for the same weight of gold, the crown would occupy a larger space than an identical one of pure gold. But how does one measure the volume of an irregular crown?

Stepping into his brimful bath, as legend tells it, Archimedes noticed water splashing over the rim. The more of him that was immersed, the more water overflowed. *Eureka!* The mundane had become momentous; to find the crown's volume, Archimedes is said to have realized, all he had to do was immerse the crown in a vessel full of water and measure the spillage. Doing so later, he informed Hieron that the crown was indeed too large for the original weight of gold. The smith was guilty. Primitive scientific deduction and measurement had one of its earliest successes. The true "gold," however, lay in Archimedes' broader conclusions; he established the key principles of buoyancy that govern the flotation of hot-air balloons, ships, and denizens of the sea. And his *Eureka!* became the joyous expletive that erupts whenever an experiment yields a sublime result or disparate ideas cohere into a beautiful theory.

This homely incident and its technical spinoff are the merest glimmer of the manifold genius of Archimedes and the profound impact he had on the development of mathematics and science. Archimedes' interests ranged widely: from square roots to irrigation devices; planetariums to the stability of ships; polyhedra to pulleys; number systems to levers; the value of the mathematical constant *pi* to the size of the universe. Yet this same cerebral man, when called upon by his king, developed machines of war so fearsome, they might have sprung from a devil's darkest imagination—weapons that held at bay the greatest army of antiquity. Ironically, it was for his feats of engineering, not for his beloved mathematics and science, that Archimedes' reputation swelled to mythic proportions in the ancient world. The Roman statesman-philosopher Cicero claimed that Archimedes possessed a genius greater than one would imagine possible for a human being.

Archimedes is universally acknowledged to have been the most proficient mathematician of antiquity and among the top mathematicians of all time, on par with the likes of Isaac Newton and Carl Friedrich Gauss. Archimedes derived the mathematical properties of parabolas, spirals, and polyhedra. He conjured geometric solids with tongue-twister names like *truncated cuboctahedron* and *rhombicosidodecahedron*, the latter an implausible sixty-two-sided conflation of abutting triangles, squares, and pentagons. He developed new ways to compute square roots, lengths of arcs, and volumes of spheres, cylinders, and cones. For the last, he used an elementary form of calculus almost two millennia before its full introduction by Newton and Leibniz. No wonder Galileo called him superhuman.

Although pure mathematics was his greatest joy, Archimedes also made seminal contributions to science. His center-of-gravity concept, now a staple of freshman physics, was among the earliest abstractions of physical objects for the purpose of analyzing nature. He solved previously intractable problems in mechanics by mathematically collapsing real objects into imaginary points of mass. Indeed, Archimedes pioneered the union of mathematics and physics that was to become a hallmark of modern scientific analysis. He is also reported to have studied optics and written a treatise on mirror reflection (now lost). And, of course, his *Eureka!* work on buoyancy was unmatched in the ancient world.

When pressed, Archimedes could be remarkably adept at invention. The hand-cranked irrigation device, commonly known as the Archimedes screw, may have been developed by him in his youth while studying at Alexandria in Egypt. There are also tantalizing reports that he built a working mechanical model of the solar system, one of the first planetariums, and designed both a steam-powered cannon and a compressed-air organ. He was also a genius in the use of levers and pulleys, boasting to King Hieron, "Give me a place to stand, and I will move the Earth." As proof of his assertion, Archimedes contrived to launch, single-handedly, a fully laden ship using what may have been a compound system of ropes and pulleys. Astounded, King Hieron proclaimed to the Syracusan citizenry, "From this day forth Archimedes is to be believed in everything he may say."

The king's declaration proved to be a harbinger of what was to come. Archimedes' treatises, rediscovered after a thousand years of collective amnesia in Europe, helped guide nascent thinkers out of the Dark Ages and into the Renaissance. Indeed, Archimedes' cumulative record of achievement—both in breadth and sophistication—places him among the exalted ranks of Aristotle, Newton, and Einstein. When contemporaries searched for a suitable honorific for the fifteenth-century architect Filippo Brunelleschi, they identified him not with his architectural forefather Vitruvius but with Archimedes. Galileo invokes Archimedes' name over a hundred times in his works. The seventeenth-century polymath Robert Hooke dubbed his innovative telescope the Archimedean Engine. And, like a highbrow graffiti artist, the teenage Isaac Newton scratched Archimedes' geometric diagrams on the wall of his school.

The name Archimedes graces computer programs, Web sites, a screw-shaped marine fossil (as well as the corkscrew in my kitchen), a short story by Aldous Huxley, an essay by Mark Twain, and, in a double dose of wisdom-by-association, Merlyn's pet owl in T. H. White's Arthurian tale *The Sword in the Stone*. Archimedes' profile is seen on the Fields Medal, the "Nobel Prize" of mathematics. And the ancient sage's tendency to become lost in cogitation may be the cultural root of characters such as Jerry Lewis's "Absent-Minded Professor" and Hubert Farnsworth in the animated television series *Futurama*.

Today there are Eureka vacuum cleaners, camping tents, and (naturally) bathtub curtains. California's motto is *Eureka*, as is the name of a prose poem by Edgar Allan Poe. Over the ages, Archimedes' vaunted lever has been exerted in the name of philosophy (René Descartes), patriotism (Thomas Paine), international relations (Thomas Jefferson), history (Honoré de Balzac), human fallibility (Franz Kafka), revolution (Leon Trotsky), political theory (Hannah Arendt), media influence (Marshall McLuhan), world peace (John F. Kennedy), and racial justice (Robert Kennedy). The Archimedean lever has likewise infiltrated literary works, such as *Don Juan*, *The Three Musketeers*, *Dracula*, and Shelley's philosophical poem "Queen Mab."

Archimedes floats through our collective consciousness: a forebear of modern ideas, a symbol of the intellectual heights to which the human species can rise. His ingenuity has inspired generations of thinkers in his own time, during the Renaissance, and up through the present day. He comes to us not merely as a name but as a flesh-and-blood character, swayed by the passions and perplexities of genius. His is a larger-than-life story, no doubt burnished by mythmakers who sought to elevate him beyond his manifold accomplishments.

Who, then, was Archimedes? If the barriers of time, language, and culture were magically breached, would Archimedes be the sprightly, avuncular intellectual—like a favorite teacher, perhaps—eager to involve us in his latest mathematical escapade? Or would he be withdrawn and cerebral, occupying some elevated plane of existence, friend only to his invisible muse? Was he roused by life's joys and tribulations or ever lost in blissful cogitation? Where does Archimedes-the-man end and Archimedes-the-myth begin? All that we have to paint the portrait of the true Archimedes are secondhand or thirdhand accounts, set down decades or centuries after his death. And, for additional clues about his personality, we have his own writings—or, at least, imperfect copies of copies of copies of his original writings. Surely, the temptation to bathe an overachiever like Archimedes in a heroic light proved irresistible to those charged with transcribing the great man's words and writing his biography.

Ancient historians are notoriously unreliable, both as reporters of events and as portrayers of character. No one can say for sure whether Archimedes truly hopped from his bath and shouted *Eureka!*—much less whether he preferred solitude over society, was an early riser, or had a predilection for oysters. Most ancient biographers embellished freely, amplifying in equal measure the affirmative aspects of those they admired and the villainous aspects of those they despised. The lives of noteworthy people were more than mere strings of occurrences; they were archetypes to be emulated or avoided. The first-century Greek biographer Plutarch, for example, cast his retrospective eye on a wide array of historical figures. His ranging account of the Roman general Marcus Claudius Marcellus indirectly illuminates Archimedes; this is how we know of Archimedes' supposed "high spirit," his fantastic war machines, his favor of abstract mathematics over "sordid and ignoble" engineering, and the purported circumstances of his death. But Plutarch was not a biographer in the modern sense as much as a moral essayist. His writings are biographical analyses of character: how events shape character, how character shapes events. To his credit, Plutarch quotes liberally from lost works of his predecessors, providing a multidimensional perspective (still woefully incomplete) on his biographical subjects.

Virtually nothing is known about Archimedes' formative years. Even the name Archimedes—literally, "Master of Thought"—may be a pseudonym bestowed upon the man by an admirer. Eutocius, who wrote about mathematics and astronomy during the sixth century A.D., refers to an Archimedean biography by one Heracleides, but that account is nowhere to be found. The twelfth-century Byzantine historian Joannes Tzetzes tells us, without evidence, that Archimedes was seventy-five years old when he died at the hands of the Romans during the siege of Syracuse in 212 B.C. That puts Archimedes' birth close to 287 B.C. Archimedes himself mentions his father, Phidias, within the text of *The Sand-Reckoner*, an expansive thought experiment on the size of the universe. Of Phidias, an astronomer, we know only that he estimated the relative sizes of the Sun and the Moon.

Whether Archimedes was raised in modest or privileged circumstances is uncertain. His father's name, Phidias, derives from that of the great sculptor of the Parthenon and was common among the artisan class. Certainly no elite family would have named a child Phidias. Cicero, a century and a half after Archimedes' death, describes his Greek predecessor as *humilem homunculum*—a humble little man—which could just as well refer to Archimedes' personal style or oratory talents as to his economic status. The first-century Roman poet Silius Italicus reports in his epic chronicle of the Second Punic War that Archimedes was *nudus opum*—destitute of means. Given Archimedes' professed devotion to his studies, mathematical accomplishment might have been all the remuneration he sought. Pursuit of the lavish life, were it even an option, would have diminished the productivity that is evident in the historical record. Presumably, Archimedes subsisted on some form of patronage, perhaps a royal retainer, in exchange for his unique civic services to Syracuse.

Plutarch intimates that Archimedes was a close friend, if not kin, of the Syracusan King Hieron II, who himself was a nobleman's son—albeit an illegitimate one. Perhaps the protomathematician and protowarrior-monarch knew each other in their youth and maintained the bond through their respective ascendancies. Or maybe their lives converged only during adulthood, in a long-running give-and-take of mutual self-preservation: Hieron's political wiles kept Syracuse at peace, while Archimedes' military inventions strengthened the city's defenses against onslaught.

A historical reference that Archimedes studied under Plato is patently false; although Plato did visit Syracuse a number of times, he died sixty years before Archimedes was born. And, no, Archimedes was not the son of the great mathematician Pythagoras, as one bygone commentator claimed; the pair were centuries and oceans apart. But there is no doubt that the young Archimedes did study mathematics with the successors of Euclid at Alexandria, in northern Egypt. (Assuming the standard chronology is reliable, Archimedes was a young man when Euclid died, but there is no mention anywhere of the two having met.)

Alexandria was founded in 332 B.C. by Alexander the Great in the wake of his campaign of conquest. By Archimedes' time, it had already grown to become the intellectual and commercial center of the Hellenistic world: a grand galaxy of buildings, monuments, wide ways, and human strivings. Along the boulevard-like Canopic Way, stretching between the Gate of the Sun and the Gate of the Moon, Alexandria's civic vigor manifested itself in spectacular Dionysian processions, one of which "included a hundred-and-eighty-foot golden phallus, two thousand golden-horned bulls, a gold statue of Alexander carried aloft by elephants, and an eighteen-foot statue of Dionysus, wearing a purple cloak and a golden crown of ivy and grapevines."

It was here, after the young Macedonian king's death, that his general, Ptolemy I Soter, established the Temple to the Muses—the Museum—and its extraordinary Library with as many as half a million documents and scrolls. (By comparison, the largest medieval European library, the Sorbonne, had less than two thousand volumes by the fourteenth century A.D.) A later regent, Ptolemy III, was an even more ardent collector. He decreed that all visitors were to relinquish any documents of literary or scientific value, which were then added to the Library's collection; in return, the visitors got cheap papyrus copies of their "donated" works. He also paid a hefty deposit to the Athenian library to borrow the state copies of works by Aeschylus, Sophocles, and

Euripides, on the premise of transcribing them; the originals never made it back to Athens. The carefully prepared Alexandrian editions of works became the exemplars whose progeny spread throughout the Hellenistic world and, eventually, to the libraries of medieval Europe.

Alexandria was a magnet to the Mediterranean region's most able intellects, including Archimedes. In addition to its rich Library, the Alexandrian Museum had research rooms, an observatory, a zoo displaying exotic species, living quarters, and a dining hall where scholars gathered to dine and debate. Here was an ancient think tank devoted to the arts and sciences, a precursor Institute for Advanced Study, whose collective scholarship became its legacy to future generations—and whose eventual decline under Christian authority in the fourth century A.D. and destruction in A.D. 642 at the hands of Islamic invaders mark one of civilization's greatest losses.

In this percolating cauldron of ideas, Archimedes met the mathematicians who would become the chief recipients and disseminators of his works—although history hints that few of these ancient practitioners were able to dissect his more advanced treatises. It was in Egypt, too, according to the Sicilian historian Diodorus, that the young Archimedes invented his eponymous screw, an inclined, hand-turned, hollow spiral, which is still used in underdeveloped regions to raise water into irrigation ditches. After some period of study at Euclid's school, Archimedes returned to his native Syracuse, where he apparently remained for the rest of this life.

Archimedes' reputation for practical creativity spread so widely, both during his lifetime and afterward, that a whole mythology of invention followed in his wake. Did Archimedes return to Egypt to oversee the construction of dikes and bridges to tame the Nile, as claimed by the thirteenth-century Egyptian historian al-Qifti? Not likely, given all indications that he spent his later life exclusively in Syracuse. Did Archimedes design naval weaponry for a mysterious Spanish monarch, Ecliderides, king of the Cilodastri? This report comes straight from one of Archimedes' most ardent latter-day admirers, Leonardo da Vinci, who claimed to have read it in a book. Yet modern historians find no record of Ecliderides or the book that allegedly contains his name.

Archimedes' everyday character—the face he presented to his fellow citizens of Syracuse—will always lie in shadow, the fertile soil of mythologizers. If Plutarch is to be believed, Archimedes exuded that otherworldly detachment so often linked with genius. Plutarch writes that Archimedes was "continually bewitched by a Siren who always accompanied him," and was "possessed by a great ecstasy and in truth a thrall to the Muses." Such was Archimedes' single-minded devotion to his mathematical ponderings that he would often "forget his food and neglect his person, to that degree that when he was occasionally carried by absolute violence to bathe or have his body anointed, he used to trace geometrical figures in the ashes of the fire, and diagrams in the oil on his body, being in a state of entire preoccupation, and, in the truest sense, divine possession with his love and delight in science." (Isaac Newton displayed similar eccentricities, leaving his meal tray untouched or inadvertently walking out of his house while lost in thought.)

Faced with the scarcity of biographical evidence, some historians have delved into Archimedes' mathematical works for clues to his character. But here, with the exception of a few personalized prefaces, they come up short. Archimedes appears to have left no written records of his vaunted engineering accomplishments. Plutarch takes this as a mark of Archimedes' disdain for anything that smacked of invention or the practical arts. He writes, "Archimedes possessed so high a spirit, so profound a soul, and such treasures of scientific knowledge, that though these inventions had now obtained him the renown of more than human sagacity, he yet would not deign to leave behind him any commentary or writing on such subjects; but, repudiating as sordid and ignoble the whole trade of engineering, and every sort of art that lends itself to mere use and profit, he placed his whole affection and ambition in those purer speculations where there can be no reference to the vulgar needs of life."

Who knows? Maybe Archimedes saw no need to chronicle his inventions, for what more concrete record would one require than the inventions themselves? So the true voice of antiquity's Master of Thought—at least as much of it as can be impressed upon the written word—is probably not to be found in hearsay reports of his feats with pulleys and levers, weaponry, or water screws.

To modern biographers, Archimedes' true character remains elusive. Seemingly out of nowhere appears this meteoric blaze of intellectual fire, leaving behind neither immediate successors nor a school to carry on his legacy. Yet if the man himself is ambiguous, his manifold accomplishments are not. Those accomplishments are all the more remarkable given the period in which he lived, an age when both civilization and the exploration of nature were in their raucous youths. Superstition and false notions invaded the study of physical phenomena, just as would-be conquerors and their armies periodically invaded Archimedes' homeland. Archimedes contemplated his mathematical problems in a time and place so riven by discord as to suffer periodic extinctions of cities. Indeed, he is said to have taken his dying breath poised between the point of a sword and a point he was pondering in mathematics. To help separate the factual from the legendary— that is, to truly walk with Archimedes—one must roll back the centuries and enter the world in which he lived.

Chapter 2

THE STORMY SEA

I believe I found the missing link between animal and civilized man. It is us.

—KONRAD LORENZ

ARCHIMEDES' STORY UNFOLDS against a tumultuous backdrop of political intrigue and assassinations, the brutal Punic Wars, Hannibal's invasion of Italy, and—tragically for Archimedes—the siege of Syracuse by the brilliant Roman general Marcellus. The Mediterranean region was ringed by competing powers, each jockeying for advantage on land and sea. Allegiances shifted frequently, as tyrants gauged each other's strengths and weaknesses; as despots subjugated the populace; as citizens overthrew their rulers; as mercenary armies demanded treasure for service; as colonists ousted indigenous peoples, and the natives pushed back. The result: almost continuous conflict.

Archimedes was born around 287 B.C. in Syracuse, an independent Greek city-state on Sicily's southeastern shore. By Archimedes' time, Syracuse had become one of the most influential and wealthy centers of the Hellenic empire. It was during this rare, decades-long respite from Sicily's incessant violence that Archimedes flourished. Had Archimedes been born a century earlier, he might have been fleeing rapacious invaders or co-opted into distracting palace intrigues; a century later, with his once-glorious Syracuse shrunken into a Roman backwater, his cerebral pursuits might have been stillborn. During most of its ancient existence, Sicily was not a place for the scholarly life. Blood spilled again and again upon the Sicilian earth in a relentless whirlwind of human savagery. Then again, the sword had a habit of intruding itself practically everywhere in the Mediterranean, and Sicily was no exception.

Like a giant triangular bull's-eye, Sicily occupies a central place in the geography of the Mediterranean. And, lying at the nexus of ancient sea routes for exploration, colonization, trade, and conquest, the island played a vital role in the early history of Western civilization. Phoenician ships from the maritime city-states of Sidon and Tyre stopped in Sicily on their twenty-four-hundred-mile voyage from the Middle East to the far Pillars of Hercules at the Strait of Gibraltar. Greek settlers established outposts along its fertile shores, displacing native dwellers inland. Etruscan pirates preyed on trading vessels in the Tyrrhenian Sea to the north. Carthaginian forces repeatedly swept the island, only to be repelled by Greek defenders and, later, by Roman legions.

Sicily's landscape and ecology have been so utterly altered by human activity that it's hard to gauge what attracted ancient travelers to its shores. The island's temperate climate, primeval forests, and once-fertile plains had proved hospitable to aboriginal Sicilians and irresistible to migrants from nearby Italy and, later, from faraway Phoenicia and Greece.

The story of Syracuse begins in 733 B.C., some five centuries before Archimedes, when Corinthian settlers drove the native populace from the mile-long, oblong island of Ortygia in southeastern Sicily. Here the founders quickly rekindled the sacred hearth fire of the goddess Hestia that had burned in their mother city. They constructed a temple to Zeus and also dedicated a Sicel spear, perhaps won in the expulsion of the natives, in the newly built sanctuary of Athena. Devotion to the gods of their homeland remained paramount in the lives of the transplanted Greeks. (Nevertheless, a hallmark of Greek science and mathematics is their strict secularity; nowhere in Archimedes' writings is there acknowledgment of any god.)

The island of Ortygia lies between a pair of sheltered, defensible harbors: the Great Harbor to the west and the Little Harbor to the northeast. A narrow channel separates the island from the mainland. With its own freshwater spring and easy access to the sea, Ortygia became the base from which the Corinthian Greeks soon spilled onto the Sicilian countryside. The displaced locals were forced to work the land for their unbidden masters, an arrangement that prompted centuries of discord between Syracusans and Sicels.

The coastal area west of Syracuse was a swamp, funneling Greek mainlanders northward onto the Epipolae Plateau, a three-and-a-half-mile-long massif that looms over Ortygia. As the Syracusans would learn centuries later, control of the Epipolae Plateau was critical to the defense of the city; whoever held the high ground had dominion over the city. The Epipolae's underlying rock includes a shallow, bowl-shaped stratum of impermeable marine clay, elevated toward the northwest; rains in the western Epipolae seep underground southeast toward Syracuse, where the waters emerge in springs. The Greeks augmented the natural water sources with a complex network of aqueduct tunnels to bring water more efficiently to the city. The easy availability of water would prove to be an asset in Archimedes' time when Syracuse endured a two-year siege by the Romans.

Other Greek strongholds arose in Sicily around this time, each one established by immigrants from a different region of Greece: to the north of Syracuse, the Chalcidian coastal cities of Naxos, Leontini, Catania, and Zancle (Messina); also to the north, Megara Hyblaea, founded by the community of Megara, near Athens; and to the west, Gela, the result of a joint expedition from Rhodes and Crete. They were joined by a number of Phoenician settlements rimming Sicily's western shore, a comfortingly short sail from their powerful sister city, Carthage, in present-day Tunisia. The island's vast interior remained largely the province of native Sicilians, with Greek and Phoenician outposts scattered throughout. So began the inevitable, yet invisible, division of territory into Greek east and Phoenician west, tempered by limited intermingling through commerce and culture. The boundary between them was mutable, but on the issue of encroachment, each side knew how far was too far and responded with force.

Although separated from their homeland, the Syracusans, along with their fellow Greek settlers, considered themselves fully Greek. They retained their language and dialect (as evidenced in the distinctly Doric character of Archimedes' writings). They also followed the established customs and religious practices of their mother city in Greece; kept a common calendar; erected buildings and monuments in the standard architectural style (although with native construction materials); laid out their cities to reflect those they had left; and carried over—at least initially—the political sensibilities of their homeland. Sicilian Greeks sent representatives to consult the oracle at Delphi, took part in religious festivals back home, and participated with great fanfare in the athletic games at Olympia. At the same time, the settlements in Sicily were entirely free of control by their respective mother cities. They were mini-states unto themselves and let their own self-interests and territorial ambitions guide their actions.

By the eighth century B.C., the migratory process that brought Archimedes' forebears to Syracuse was already well under way. The initial impetus behind Greek migrations overseas was not one of empire but of commerce. Coastal outposts—emporiums—were established throughout the eastern Mediterranean, where Greek wine and oil were exchanged for metals, fabrics, and slaves. This elective dispersal of Greek citizens from the mainland was accompanied by a second migratory wave, this one of brute necessity. Greece had evolved into a multitude of independent city-states, isolated from one another by the country's rugged topography. The mountains likewise precluded large-scale farming, making it increasingly difficult for communities to sustain their burgeoning populations. Inevitably, Greek cities looked to disperse their surplus citizenry to the Aegean islands, Asia Minor, North Africa, western Europe, and Sicily. (Athens sustained its population of several hundred thousand only by importing food.)

To most ancient Greeks, the idea of leaving their home city— the polis—was anathema, a fate typically reserved for miscreants or bastard sons (as in the colonization of Taras—modern Taranto— by the Spartans in 706 B.C.). Aristotle viewed the formation of the Greek polis as inevitable: "Humans are beings who by nature live in a *polis*." Aristotle adds, with a hint of sarcasm, that those who live outside such a social structure must be either beasts or gods. For a member of the polis to be shipped off to a foreign land meant a severing from home and family forever. Promise of adventure or commercial opportunity probably did little to assuage the trepidation of most occupants forced to leave the polis. Greek cities were desperate to relieve their swelling economic burden. Somebody had to go. And the selection process was frequently draconian.

One ancient inscription describes how the citizenry of the south Aegean island of Thera—modern Santorini—chose who would sail to the new colony of Cyrene in North Africa: "One adult son [from each family] is to be conscripted . . . If any man does not wish to go when the polis sends him, he shall be liable to the death penalty and his property shall belong to the [people]. And the man harboring or concealing him, whether he be a father [aiding his] son or a brother his brother, is to suffer the same penalty as the man who does not wish to sail."

Excavate an ancient colonial Greek cemetery, such as the Fusco cemetery west of Syracuse, and you'll discover two things: Women were interred there from the colony's earliest years; and scattered among the bones are ceramics, jewelry, and ornamental items of both aboriginal and Greek origin. In Syracuse, as in most Greek settlements, there appears to have been a rapid mingling of cultures. Upon arrival in their new homeland, Greek men—for, with the exception of a priestess or two, the initial colonizing party was exclusively male—probably sought wives among the native populace, instead of waiting for Greek women to arrive from the mother city. In the heyday of Greek colonization, centuries before marriage to "barbarians" was actively discouraged, only the father's parentage determined the "Greekness" of the children. It was almost certainly permitted, and maybe expected, that colonists would marry women from indigenous tribes. Indeed, there was plenty of precedent for intermarriage in the poetic fiction of Greek historical legends, including the *Odyssey*, that pervaded the culture at the time.

To what degree these interracial marriages were coerced isn't clear; the practice probably varied from colony to colony, if not from woman to woman. Aristotle records that, in the founding of Massalia by Phokaian Greeks, the daughter of the local Celtic king spurned her native prospects and chose a Greek court guest to become her husband. Likewise, there is ample evidence of voluntary intermarriage between Greek men and Libyan women during the establishment of Cyrene. On the other hand, the historian Herodotus describes the founding by Ionian Greeks of Miletus in Asia Minor, where the Greeks slew all the native men, then married their wives and daughters. In protest, it is said, the women swore an oath never to eat meals with their husbands or call them by name, a practice that went on for centuries.

By custom, the original colonists claimed and subdivided the proximate territory among themselves, becoming the community's landed gentry. Subsequent immigrants received their own parcels of land until there was no more in the immediate area, and the entry gate to the aristocratic classes effectively swung shut. It's no surprise that the default form of government in many colonies was an oligarchy of landowners. Civic unrest or security issues frequently led to the overthrow of the oligarchies, as military-backed tyrants promised colonists better governance and protection. Democratic governments never lasted very long in Sicily.

In his *Laws*, Plato describes how land might have been equitably divided among a colony's settlers. First, the polis, or city center, is established, within which is built a walled sacred area—the acropolis—designated for the worship of Hestia, Zeus, and Athena. The remaining polis area is divided into twelve equitable parts, more desirable lots being smaller than the less desirable ones. Next, a large number of residential lots are marked out; for example, Plato specifies 5,040 lots (that is, the product $1 \times 2 \times 3 \times 4 \times 5 \times 6 \times 7$). Each of these lots is divided into two sublots, whose acreage again is based on quality of the land. Every sublot near the polis is paired with a sublot farther out, and these pairs are distributed equitably among the colonists, along with two houses each. "And thus," concludes Plato, "the foundation is complete."

Although Plato stipulates 5,040 lots, he adds, "Every legislator ought to know so much arithmetic as to be able to tell what number is most likely to be useful to all cities . . . for no single instrument of youthful education has such mighty power, both as regards domestic economy and politics, and in the arts, as the study of arithmetic." The goal of his methodology, Plato explains, is to "banish meanness and covetousness from the souls of men, so that they can use them properly and to their own good." In this, Plato failed—not surprisingly.

Ever present in the minds of the Sicilian Greeks was the specter of invasion, whether by their arch-nemesis Carthage, whose armies periodically trammeled the island; by Greek armadas from the mother country; by their fellow Greeks in neighboring cities; or by indigenous tribes from Sicily's interior. The near-constant threat of marauders laid fertile ground for the rise of military tyrants. During brief and infrequent periods of peace, democracy sometimes flourished, only to be abandoned when aggressors, real or invoked, appeared on the horizon.

The first Greek-Sicilian tyrant of significant stature was Hippocrates (not to be confused with the famous physician of the same name), who rose to power during the fifth century B.C. in Gela, some fifty miles west of Syracuse. With his large mercenary army, Hippocrates brought under his control the major Greek settlements in eastern Sicily: Naxos, Zancle, and Leontini. Syracuse escaped for a time through mediation until Hippocrates' cavalry commander and eventual successor, Gelon, took advantage of factional strife and seized the city. (Gelon won the chariot competition at Olympia in 488 B.C.) Eager to leave the hinterland for the big city, Gelon installed himself as Syracusan ruler, consolidating his power by importing to Syracuse half of Gela's population, citizens and mercenaries alike.

Subsequent years saw an enormous buildup of the Syracusan army and navy. If ancient reports are to be believed, the fighting force consisted of some two hundred warships, twenty thousand infantry, and five thousand cavalry, archers, and slingers. In 480 B.C., Gelon's war machine routed Carthaginian invaders at Himera in northwestern Sicily. Six years later, Gelon's brother and successor, Hieron, defeated the Etruscans at Cuma in Italy. With these blows, and with a massive tribute exacted from Carthage, Syracuse not only enjoyed hegemony over all of Sicily but became the reigning Mediterranean superpower. (Both Sparta and Athens had asked Gelon for military assistance to repel the Persians, but refused Gelon's condition that he lead their combined forces into battle. Their victory at Salamis, coupled with the near-simultaneous victory of Syracuse over Carthage, was held up as evidence of Grecian superiority over the "barbarians.")

Although Gelon and afterward Hieron brought great wealth to Syracuse, the citizens had grown weary of the tyrants' predilection for war, land reallocation, and mass population transfers. In 466 B.C., the Syracusans drove Hieron's successor, Trasibulo, and thousands of mercenary-settlers from the city and established a democratic government—at least in name. Despite the presence of a popular assembly that passed laws and decided foreign policy, political and military leaders were still drawn from the wealthy elite, and class division and distrust remained as entrenched as ever. The lure away from Athenian-style democracy and once again toward tyranny came, paradoxically, from Athens itself.

Since the early 450s, even while actively engaged against Sparta, Athens had shown periodic interest in expanding its influence into Sicily and southern Italy. To this end, Athens formed alliances with a number of overseas cities, including Leontini, which lay practically on Syracuse's doorstep. The alliance was probably initiated by the citizens of Leontini, who were justifiably alarmed by the growing might of Syracuse. When Leontini and Syracuse went to war in 427 B.C.—and dragged their respective Sicilian allies into the fray— Leontini called on Athens for help. Athens responded with a naval force but waged a desultory campaign and let the Sicilian Greeks duke it out among themselves.

At a peace conference in Gela in 424 B.C., the Syracusan politician and general Hermocrates warned his fellow Sicilians of the Athenians' broader stratagem: "It is Sicily as a whole, in my judgment, that the Athenians are scheming to take over. And it is the Athenians also, far more than any words of mine, who ought to be making us come together and settle our differences. They have the greatest power of any Greeks and they have come here with a few ships to wait until we make mistakes; in the name of a normal, lawful alliance, they are turning our natural hostility toward each other to their own profit . . . [W]hen they realize that we are worn out they will one day come here with a larger force and try to bring everything here under their control." A truce was concluded among the warring Sicilians, and the Athenian fleet left.

Hermocrates' words proved prophetic. In 415 B.C., the Athenians were back, this time with the largest armada ever assembled by a Greek city: 250 ships and twenty-five thousand soldiers. But the Athenians frittered away any advantage they might have had by seeking alliances among Syracuse's neighbors and by raiding settlements for treasure. Their eventual destruction at the hands of the Syracusans, led by the Spartan commander Gylippos, was one of the signal events in Greek history—"an utter calamity," in the eyes of the ancient historian Thucydides, "army, fleet, absolutely everything was destroyed." Some seven thousand Athenian combatants were sent to the stone quarries or branded on the forehead and sold into slavery. (According to Plutarch, the few who were released were those who could recite by heart the choruses of Euripides, whose works were esteemed by the highly educated Syracusans.) In a supreme irony, the Syracusan assembly—with military backing—instituted reforms that took their democratic government closer to the ideal of the vanquished Athenian counterparts. It was not to last.

In 406 B.C., following another Carthaginian onslaught on Sicily, a twenty-four-year-old military man named Dionysius seized command of Syracuse. Quick to take advantage of long-time class resentment against the rich, Dionysius discredited his political rivals before imprisoning them in a limestone cave in the hills. (In 1586, the Renaissance artist Caravaggio, fleeing a murder warrant, hid in the prison's winding chamber; he dubbed it the Ear of Dionysius for its shape and peculiar echoic properties.) Once in power, Dionysius expelled the residents of the Syracusan central city of Ortygia onto the mainland and turned the island into a redoubt for himself and his mercenary army. He is said to have stripped the solid-gold robe from the statue of Zeus, re-marking that the gleaming cover was "too cold for winter and too hot for summer." According to legend, Dionysius arranged a ban-quet at which a nettlesome sycophant, Damocles, was forced to dine underneath a sword suspended from the ceiling by a horse hair; thus, Dionysius explained to the trembling Damocles, is the insecurity of royal office.

In 405 B.C., Dionysius signed a duplicitous nonaggression pact with the Carthaginians, ceding control of much of Sicily, yet fully intending to attack their settlements at a more opportune time. Meanwhile, he made preparations for both the defense of his city and the taking of others. Dionysius read a lesson from the Athenian siege of Syracuse decades earlier: Had it not been for the timely arrival of reinforcements from Sparta, the Atheni-ans would surely have starved the city into submission. A similar lesson lay in the recent Carthaginian assaults on Selinus and Himera, both of which had fallen victim to battering rams, wheeled siege towers, and, in their wake, Carthaginian atrocities. Dionysius had seen for himself the Carthaginian juggernaut as it pummeled the walls of Syracuse's sister city, Gela. The sight had convinced him to flee under cover of darkness to Syracuse with the entire Gelan population in tow. Before he dared invite the retribution of Carthage by attacking its allied cities on Sicily,

Dionysius would turn Syracuse into an impregnable fortress. And that meant securing the high ground of the neighboring Epipolae Plateau.

The Epipolae already had the makings of a defensive bulwark. Its steep sides made direct approach difficult, except at a few places. Appealing to Hellenic chauvinism—and commandeering where cajoling failed—Dionysius marshaled the collective labor of sixty thousand workers to effectively heighten the plateau by erecting a stone wall atop its entire sixteen-mile periphery. The historian Diodorus describes the crash project as a marvel of efficiency: six thousand yoke of oxen to transport the four-foot-long stone blocks; labor crews of two hundred, each assigned a one-hundred-foot section of wall; teams of master builders to supervise construction; prizes to the fastest crews. Dionyius himself reportedly joined in the labor, to the sure surprise of the peasants around him. The wall went up in just three years and incorporated not just the plateau but the harborside and city center as well.

Despite all this construction, the Epipolae remained vulnerable at its western extreme, where the inland plain merges via a narrow land bridge onto the otherwise raised plateau. An aggressor army could sweep in across this topographic Achilles' heel and wheel its siege engines right up to the western gate. Surveying the scene, Dionysius saw opportunity in the form of a nondescript hillock three hundred feet south of the gate called Euryalus, meaning "wide nail" or "wart." On this mound, whose southern side fell away in a steep slope and whose northern side overlooked the vulnerable approach to the Epipolae, he erected a fortress. Bristling with defensive battlements, Euryalus would evolve over the years to take full advantage of a new artillery weapon developed by Dionysius's engineers: the catapult. The final phase in the fortress's evolution would be guided by a later son of Syracuse: Archimedes.

In 398 B.C., the tireless Dionysius launched his offensive against the Carthaginian settlements of western Sicily. Carthage counterattacked by both land and sea, to little effect; the new fortifications around Syracuse rendered direct assault almost impossible. Although victorious in the end, Dionysius built no unified nation in his thirty-eight years at Sicily's helm. Instead, he pursued policies that served to preserve his power, enrich his supporters, and turn back repeated retaliation by Carthage. He had provided Syracuse with the most advanced defensive battlements in the ancient world but also left a decades-long period of instability in Syracusan affairs.

The political situation in Syracuse grew so unstable that Plato himself was twice summoned, in 367 and 361 B.C., to knock some philosophic sense into the head of the slothful Dionysius II, who had succeeded his father. Plato was unimpressed with the youthful tyrant and with his subjects: "Where happiness derives from gorging oneself at table twice a day as the . . . Syracusans do . . . and never sleeping alone at night, such habits having been followed from youth, a man never has the chance to grow wise." For his troubles, Plato was placed under house arrest for alleged collusion against the monarch. He was released after a few months and returned to Athens in utter disgust.

After Dionysius II came a succession of Syracusan tyrants, each one murdered or driven out after a brief reign. Sicily descended into chaos, with wealthy aristocrats pitted against champions of the commoners who pressed for more equitable land distribution, and with mercenaries providing the muscle to advantage one side or the other. While Syracuse was preoccupied with its own internal politics, Carthage crept back into western Sicily.

Facing utter collapse, Syracuse defied centuries of tradition and turned to its mother city, Corinth, for help. Corinth, once the foremost shipbuilding center in archaic Greece, had fallen on hard times. Yet the loose ties of kinship and the renewed threat from Carthage must have swayed opinion, for it responded favorably to its daughter city's plight. Help arrived in 344 B.C. in the form of a small army led by a less-than-heroic figure: an elderly recluse named Timoleon, whose most notable contribution to Corinthian affairs had been the assassination of his older brother, the brutal tyrant Timophanes. Through skillful diplomacy and a ruthless military campaign against the Carthaginians, the unlikely commander brought all of Sicily back under Syracusan control.

Timoleon had a vision for Sicily's future, and his reign was all that previous tyrants' were not. He ordered the immediate destruction of Dionysius's fortified palace in central Syracuse and instituted democratic reform. He made peace with the native Sicilian tribes. And with his promise of free land, some sixty thousand settlers arrived at Syracuse from Greece, Italy, and greater Sicily, replenishing the city's economic base and boosting agricultural output. Education was seen as vital to Sicily's future. "So many were employed in the teaching of the youth . . . ," remarked one second-century wag, "that it became a proverb, He is either dead or is teaching letters."

Herman Melville described Timoleon's impact on Sicily in verse:

On Sicily's fields, through arduous wars,
A peace he won whose rainbow spanned
The isle redeemed; and he was hailed
Deliverer of that fair colonial land.
And Corinth clapt: Absolved, and more!
Justice in long arrears is thine:
Not slayer of thy brother, no,
But savior of the state, Jove's soldier, man
divine.

Prosperity continued after Timoleon's death around 336 B.C., although democracy did not. A new tyrant, Agathocles, promoting himself as the champion of the masses, ruled with an iron fist up to his assassination in 289 B.C., just before Archimedes' birth. Disharmony returned to Syracuse until Hieron II ascended to power two decades later, when Archimedes was probably showing early signs of genius during his adolescence. The illegitimate son of a Syracusan nobleman, Hieron had served under Pyrrhus, the Greek adventurer-king, whose costly triumph over the Romans in 280 B.C. originated the term *pyrrhic victory*. Rising to commander in chief of Pyrrhus's army, Hieron seized power in his native Syracuse in 270 B.C. When his own veteran mercenary troops threatened to rebel, Hieron sent them into battle against the Campanians at Massana, holding back his Syracusan reserves while his mercenaries were cut down. Five years later, at the head of a restored army, Hieron defeated the piratical group known as the Mamertines— the "sons of Mars"—and was proclaimed king of Syracuse.

Once ensconced in Syracuse, Hieron abandoned further territorial ambitions. Instead, he preserved his compact empire by threading a complex diplomatic gauntlet between superpowers

Rome and Carthage, each of which had designs on Sicily's riches. Through a pragmatic jig of shifting alliances and treaties, Hieron II brought Syracuse an uncharacteristic, six-decade-long period of stability. It was during these years that Syracuse's foremost citizen grew up and prospered. While King Hieron held disruptive forces at bay, Archimedes focused his attentions on science and mathematics.

Chapter 3

EUCLIDEAN FANTASIES

An independent world,
Created out of pure intelligence.

—WILLIAM WORDSWORTH, *THE PRELUDE,*
BOOK 6, *CAMBRIDGE AND THE ALPS*

ARCHIMEDES' MATHEMATICAL EDIFICE rests, in part, on a series of sophisticated treatises about the geometric and, in some cases, physical properties of figures bounded by lines, curves, or surfaces: *On the Sphere and Cylinder, Measurement of a Circle, On Conoids and Spheroids, On Spirals, On the Equilibrium of Planes, Quadrature of the Parabola, On Floating Bodies,* and the *Method of Mechanical Theorems.* In these works, he devises the means to compute all manner of quantities, such as the size of the right triangle whose area is equal to that of a given circle; the radius of the circle whose area is equal to that of a specified cone; the area bounded by a section of a parabola or a given portion of a spiral; the center of gravity (balance point) of various plane figures; the ratios of various segments generated by intersecting circles and lines; the surface area and volume of a sphere.

In his breathtaking variety of geometric ponderings, Archimedes seems a mathematician at play, romping through Euclidean fields with abandon. He frequently overleaps mundane forms—lines and planes—in favor of arcs and curved surfaces, the thorniest shapes to analyze. From the Euclidean foundation laid by his predecessors, Archimedes scales the heights of complexity as a rock climber ascends a cliff, patiently and incrementally, elevating himself with geometric handholds overlooked by the lesser skilled.

Plutarch tells us that Archimedes believed his greatest achievement to be the discovery of a curious numerical relationship between spheres and cylinders. In particular, Archimedes envisions a sphere tucked snugly within a given cylinder. The walls of the circumscribed cylinder touch the sphere all around its circumference, and the cylinder's top and bottom likewise touch the top and bottom of the sphere. (Think of a basketball in a custom-fitting waste can.) Through a maze of mathematical logic, Archimedes computed the surface area of each figure—the square-footage of its curved exterior if cut and laid flat like a map. To his surprise, the surface area of the sphere invariably worked out to be precisely two-thirds the surface area of the circumscribed cylinder. It didn't matter how small or large the figures were; as long as the sphere just fits within its enclosing cylinder, the ratio of their surfaces is always a starkly simple 2:3.

Archimedes also proved that the 2:3 ratio holds as well for the relative volumes of the sphere and cylinder. If the sphere's interior space has a two-cup capacity, then the circumscribed cylinder holds precisely three cups. This numerical correspondence might seem almost trivial, even to the mathematics enthusiast, but to Archimedes, it was a revelation. Here was evidence of the mathematical design of the world, profound in its simplicity and utter constancy. So significant was this finding to Archimedes that he instructed his relatives to carve a nested sphere and cylinder on his tombstone. (In 75 B.C., well over a century after Archimedes' death, Cicero identified the once-famous Syracusan's overgrown grave site by the unique monument.)

Early Arabic mathematicians credit Archimedes with an elaborate geometric technique to create a regular heptagon, a seven-sided figure deemed impossible to draw in the classical Euclidean way with only compass and straightedge. No doubt impressed, the tenth-century geometer Abu Sahl al-Kuhi refers to him in a letter as the "imam of mathematics."

In the *Method*, Archimedes computes areas and volumes of figures by applying a precursor form of reckoning akin to integration, millennia before Isaac Newton and Gottfried Leibniz defined it as an essential element of calculus. Here, too, Archimedes utilizes practical physics—the balancing of objects of various shapes—to foster the solution of abstract mathematical problems. Archimedes' physical analyses of buoyancy, flotation, and stability are also supremely mathematical in character. In addition to the extant treatises, there is evidence of lost mathematical studies of polyhedra, number systems, mechanics, and mirror reflections.

Archimedes was also a master of computation. When a competitor reputedly criticized his work on the multiplication of large numbers, Archimedes challenged him with a fiendishly convoluted mathematical brain teaser, popularly known as the Cattle Problem, which was not fully solved until the computer age. In *The Sand-Reckoner*, his mathematical entertainment for the ruler of his native Syracuse, Archimedes tallies up how many grains of sand it would take to fill the universe, inventing a shorthand number system to convey the answer. That Archimedes was able to accomplish such mathematical feats as these before the introduction of trigonometry, algebra, and even the decimal number system, makes them all the more remarkable.

Plutarch writes of Archimedes' treatises that it "is not possible to find in all geometry more difficult and intricate questions, or more simple and lucid explanations." That being said, most of Archimedes' treatises present a highly edited version of his thought process, with blind alleys and procedural details swept away. The logical path of proof proceeds sequentially, unerringly, to its intended destination, as though Archimedes were following some mathematical GPS of the mind. What's missing is any indication as to how Archimedes determined the proper steps of the proof in the first place. Was his process mathematical trial and error? Or was it more the forward-looking, half-intuitive approach of a chess master? Only in the twentieth century, with the chance discovery of a unique explanatory treatise, did historians gain insight into Archimedes' thought process and working methods.

In his collection of Archimedes' works, the noted mathematics historian T. L. Heath describes the ancient sage's particular appeal to the analytically minded: "One feature which will probably most impress the mathematician accustomed to the rapidity and directness secured by the generality of modern methods is the *deliberation* with which Archimedes approaches the solution of any one of his main problems. Yet this very characteristic, with its incidental effects, is calculated to excite the more admiration because the method suggests the tactics of some great strategist who foresees everything, eliminates everything not immediately conducive to the execution of his plan, masters every position in its order, and then suddenly (when the very elaboration of the scheme has almost obscured, in the mind of the spectator, its ultimate object) strikes the final blow. Thus we read in Archimedes proposition after proposition the bearing of which is not immediately obvious but which we find infallibly used later on; and we are led on by such easy stages that the difficulty of the original problem, as presented at the outset, is scarcely appreciated."

Heath describes Archimedes' treatises as "without exception, monuments of mathematical exposition; the gradual revelation of the plan of attack, the masterly ordering of the propositions, the stern elimination of everything not immediately relevant to the purpose, the finish of the whole, are so impressive in their perfection as to create a feeling akin to awe in the mind of the reader." Reflecting on the seemingly effortless virtuosity of Archimedean treatises, the Stanford classics professor Reviel Netz characterizes their execution as "inventive to the point of playfulness." To Heath, Archimedes is Beethoven; to Netz, he is Mozart.

Archimedes' playfulness may likewise be evident in the way he "published" his findings. There were no periodic journals in the ancient world, as there are today, where researchers, through timely publication, claim ownership of an original finding. From his home base in Syracuse, Archimedes announced his mathematical breakthroughs in the form of letters to faraway correspondents. These letters have come down to us as introductions to the various treatises. Reading the letters today opens a window onto Archimedes' professional character and his relationship with his compatriots. At times respectful, at times provocative, Archimedes clearly had his favorites, uppermost his former Alexandrian colleague, the mathematician and astronomer Conon.

Like a Hollywood movie trailer, Archimedes' letters often merely announce a mathematical breakthrough and instruct (dare?) his colleagues to prove the proposition for themselves. Evidently,

his correspondents begged Archimedes for solutions, either because they were stumped or because they wished to preserve for
the ages the master's own procedure. Archimedes frequently com-
plied, but only after considerable delay. Archimedes obliquely nee-
dles one young mathematician, Dositheus, that if Conon were still
alive, *he* would be able to solve the problems.

In the letter that accompanied his treatise *On Spirals*, Archimedes
issues a stern warning to would-be plagiarists: Two of his previously
sent propositions are, in fact, false, so that "those who claim to discover everything, but produce no proofs of the same, may be con-
futed as having pretended to discover the impossible."

In the decades before Archimedes arrived in Alexandria, Euclid
compiled and extended the fundamental geometric postulates of
his predecessors. The organization and stepwise narrative momentum of Euclid's proofs became the standard against which subsequent mathematicians, including Archimedes, measured themselves.
Euclid's masterpiece, *The Elements*, would become the greatest
mathematical textbook of all time and one of the most widely
translated and published works ever. In thirteen books (chapters, in
the modern sense), Euclid lays out a comprehensive survey of geometric proofs about straight-sided constructions—lines, triangles,
squares, and the like; properties of numbers, such as divisibility of a
number by other numbers; and finally, ratcheting up the complexity, properties of circles, spheres, and polyhedra.

Archimedes' treatises frequently pick up where Euclid and his
predecessors leave off, these earlier discoveries providing the
springboard from which Archimedes launches into uncharted territory. Archimedean treatises follow the rigidly logical structure of
Euclid's *Elements*. First come definitions, followed by axioms—or

common notions, as Euclid calls them—such as *things which equal the same thing also equal one another* or *the whole is greater than the part.* Next come previously proven postulates regarding the construction and properties of lines and plane figures, and finally the geometric proof itself.

Unlike modern mathematical writing, which is condensed into the shorthand of symbols, equations, and diagrams, Archimedes and his contemporaries present their analyses in prose. Here, words bear the full burden of conveying complex, sometimes abstract, meaning. Verbose as these mathematical essays might seem by modern standards, they nonetheless express arguments with supreme clarity. This ancient Greek prose style was standardized to a large extent, yet Archimedes' particular Doric dialect rings clear and often helps distinguish his voice from those of others.

Archimedes' research was complicated by primitive Greek concepts and notation conventions about numbers. Greeks viewed numbers not as abstract measures of quantity but as a language element inseparable from a collection of objects. It's as though the adjective *fast* possessed no meaning until mated with a noun, as in *fast runner.* Likewise, in the phrase *ten lords a-leaping,* the number *ten* would have had no independent, out-of-context meaning. And the ancient Greeks had no number *zero.* Since every number, by definition, was tied to some physical object, a collection with a count of *zero* would have been patently absurd.

Although fundamentally geometric in nature, Archimedes' treatises nonetheless involve extensive arithmetic calculations with numbers that range from fractions to many millions. He states the square root of 3 to great precision (without revealing how he had computed it) and also tallies up long numerical progressions like the sum of the squares of many whole numbers, $1^2 + 2^2 + 3^2 + 4^2 + \ldots$, and the sum of the powers of a fraction, such as $1 + \frac{1}{4} + (\frac{1}{4})^2 + (\frac{1}{4})^3 + (\frac{1}{4})^4 + \ldots$.

To perform heavy-duty arithmetic, Archimedes probably used a common counting board, precursor of the familiar rod-and-bead abacus. Archimedes' counting board might have been either a tray filled with fine sand—the Greek word for counting board, *abax*, derives from the Hebrew *'abaq*, for dust—or a slab of wood or marble with a series of marked columns. Into these columns he arranged counting beads. In a pinch, he might have scratched a virtual counting board into the earth and shifted pebbles among its furrows.

Figure 3-1 shows a counting board with columns indicating the place values that make up a number. In this case, there are columns for the *thousands* place, *hundreds* place, *tens* place, and *ones* place. Here, the number 32 is recorded by placing 3 beads in the *tens* column and 2 beads in the *ones* column, that is, 3 *tens* plus 2 *ones* totals up to 32.

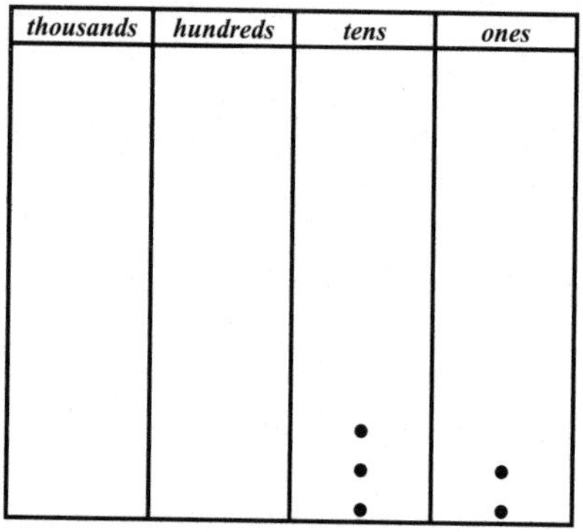

thousands	hundreds	tens	ones
		•	
		•	•
		•	•

Figure 3-1. Counting board representation of the number 32

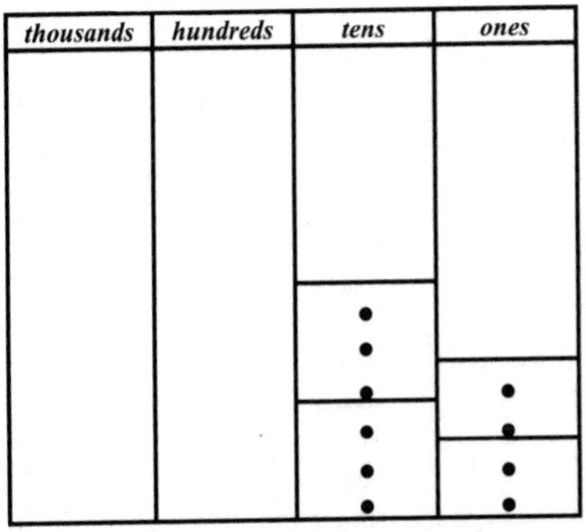

Figure 3-2. Counting board representation that the sum 32+32 or,
equivalently, the product 32×2 equals 64

Figure 3-2 shows how Archimedes would have recorded the sum
32 + 32, or equivalently, the product 32×2. The beads already on the
board for the first number 32 are joined by a like set representing
the second number 32. The result is 6 beads in the *tens* column and
4 beads in the *ones* column, representing the number 64. (For clarity, a few horizontal guide lines are indicated.)

In figure 3-3, Archimedes computes the sum 32 + 32 + 32 + 32,

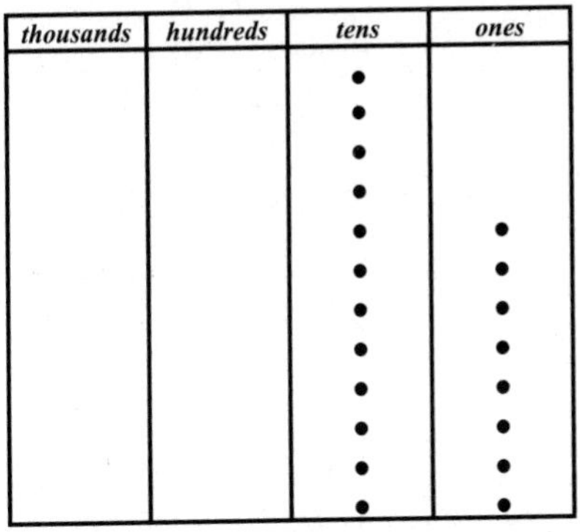

Figure 3-3. Counting board representation of the product 32×4 before
correcting the overflow in the tens column

or equivalently, 32×4. By successive additions, he winds up with 12 beads in the *tens* column and 8 beads in the *ones* column. But any time there are 10 or more beads in a column, the equivalent value of the 10 beads is carried over to the next column. For example, 10 beads in the *tens* column indicates a value of 10×10, or 100; the 10 beads in the *tens* column can be replaced by 1 bead in the *hundreds* column. If there are 12 beads in the *tens* column, as Archimedes has here, he would replace 10 of the 12 *tens* beads with a single *hundreds* bead, leaving 2 *tens* beads behind. His final configuration for the product 32×4 is shown in figure 3-4: 1 bead in the *hundreds* column, 2 in the *tens* column, and 8 in the *ones* column, that is, 128. Of course, Archimedes would have done such a computation in his head and reserved the counting board for more complex arithmetic.

While Archimedes' arithmetic would have followed the same place-value rules as we do today, not so with the recording of numbers. Writing down numbers in our modern decimal system

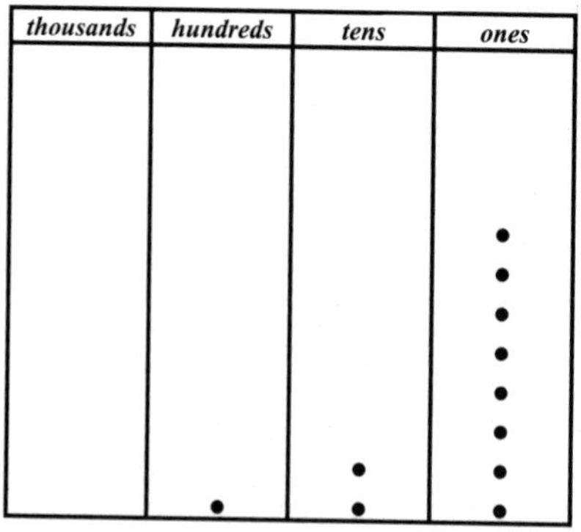

Figure 3-4. Counting board representation of the product 32×4 after correcting the overflow in the tens column

is straightforward. Only ten symbols are required, one to represent each of the digits from zero through nine. (It's no coincidence that the word *digit* is rooted in the Latin *digitus*, for finger or toe.) The introduction of the concept of zero by mathematicians in India during the seventh century provided a convenient way to indicate an empty place in a multidigit number. There is no need to create symbols for multidigit numbers, because these are written as various combinations of the ten basic digit symbols. Thus, 40 is cobbled together from a 4 in the *tens* place and a 0 in the *ones* place. The number zero allows us to provide a clear distinction between, say, 40 and 400, or between 404 and 440, a feature the ancient Greek system lacked.

From the modern perspective, the writing and reading of numbers in the ancient Greek system might have made a suitable thirteenth labor for Hercules. The Greeks—Archimedes included—represented numbers as letters from their alphabet. The number 1 is the letter α (alpha), 2 is β (beta), 3 is γ (gamma), and so on. But instead of recycling and combining the basic symbols when forming numbers larger than 9, as we do today, the Greeks assigned selected two- and three-digit numbers their own unique letter designations: 10 is ι (iota), 20 is κ (kappa), 30 is λ (lambda), and so on up to 90, which is the archaic letter ϙ (koppa); 100 is ρ (rho), 200 is σ (sigma), 300 is τ (tau), and so on up to 900, which is the archaic letter ϡ (san). The entire list of Greek number-letter correspondences is shown in table 3-1. Archimedes would have written the number 164 as the letter combination ρξδ; and with no Greek zero, the number 105 would have simply been ρε. Numbers between 1,000 and 9,999 also used the standard letters but were flagged by a special punctuation mark, while numbers larger

Table 3-1. Table of ancient Greek number-letter correspondences

1	α	alpha	10	ι	iota	100	ρ	rho
2	β	beta	20	κ	kappa	200	σ	sigma
3	γ	gamma	30	λ	lambda	300	τ	tau
4	δ	delta	40	μ	mu	400	υ	upsilon
5	ε	epsilon	50	ν	nu	500	φ	phi
6	ς	digamma★	60	ξ	xi	600	χ	chi
7	ζ	zeta	70	ο	omicron	700	ψ	psi
8	η	eta	80	π	pi	800	ω	omega
9	θ	theta	90	ϙ	koppa★	900	⟩	san★

★ archaic letters

than 10,000 were indicated by the letter *M*, from the Greek for *myriads*.

Archimedes' treatise *Measurement of a Circle* contains one of his most accessible mathematical achievements: the determination of the numerical constant π. Ancient geometers were fascinated with the circle. With its supreme symmetry and uniform contour, the circle was deemed the most sublime of all geometric figures. For instance, only the circle and its three-dimensional cousin, the sphere, embodied an aesthetic quality worthy of celestial bodies and the orbits along which they moved. The circle is the only two-dimensional figure accorded its own fundamental postulate in Euclid's geometry. (*Given any straight line segment, a circle can be drawn having the segment as radius and one endpoint as center.*) In the fifth century B.C., the Greek philosopher-statesman Empedocles likened the nature of God to a circle whose "center is everywhere and the circumference is nowhere." And in Aristophanes' play *Birds*, from 414 B.C., the buffoonish geometer Meton alludes to efforts by mathematicians to figure out how to square the circle, that is, to compute its area. The Greek expression *circle-squarer* came to refer to someone who attempts the impossible.

It was already well known in Archimedes' day that the ratio of a circle's circumference to its diameter is always the same number—slightly more than 3—regardless of the circle's size. Only in 1706 did the Welsh mathematician William Jones (who made his living by lecturing about mathematics in London coffeehouses) designate this cardinal ratio by the Greek letter π, by which it has been known ever since. Nowadays, most school kids can rattle off π to a few decimal places: 3.14. To number mavens or anyone whose calculator has a π key, it's 3.1415926. In fact, the decimal representation of π has no end, and its ceaseless array of numbers contains no repeated sequences. In 1999, mathematicians at the University of Tokyo used a computer program to calculate π to 206,158,430,000 decimal places. The number π pops up frequently in analyses of natural phenomena—gravity, orbits, electric fields of subatomic particles—since the geometry of trajectories and physical forces is so often inherently circular, elliptical, or spherical.

The earliest attempts to deduce π were conducted by direct measurement. It was easy to draw a circle if one wasn't too picky about its appearance. Some ancient geometer must have fixed one end of a horizontal wooden rod to a wax tablet, secured a vertical stylus to the rod's other end, and rotated the stylus around the fixed end to trace out a circle. The length of the rod is the circle's radius; twice the rod-length is the circle's diameter. Counting how many of these diameter-lengths would fit end-to-end around the circumference—using, say, a flexible, diameter-length reed—would have yielded an approximation of π. The biblical passage I Kings

7:23, about the construction of Solomon's temple, pegs the ratio loosely at 3. Egyptian geometers were more precise: An analysis of a circular plot of land from around 1650 B.C. gave π as the fraction 256/81, which is equivalent to about 3.160. Ancient Babylonians from the same era adopted the fraction 25/8, about 3.125.

While Archimedes had ample occasion to inscribe circles into wax or into the dirt at his feet, he never would have reckoned π through crude measurement. Such a manual procedure would have chafed against his mathematical sensibilities. And as careful as he might have been, his π would have been, at best, only marginally better than his predecessors'. To a Euclidean geometer like Archimedes, computing π from measurement of a physical circle was doubly flawed. Not only did the process lack precision; it also lacked the refinement and certainty of pure mathematical reasoning. A Euclidean-style proof was *proof*, in its strictest sense, of a mathematical assertion. Anything else was merely illuminating, instructive, a way station on the road to proof. Archimedes modified a process developed by his predecessor, Eudoxus, and blazed a theoretical path to π.

Archimedes could see it hovering before his mind's eye: a circle—free of any distortions, perfect in every sense, scribed into his consciousness by the radius of pure imagination. Why construct and measure a crude physical circle when a circle of the mind might, in theory, be measured with far greater precision by purely geometric means? And if the illusive circumference, by virtue of its continuous curvature, itself defies measurement, then measure instead the perimeter of a straight-sided figure—a polygon—that stands in for the circle. At first glance, such a technique might appear ridiculous: The simplest polygon, a triangle, bears virtually no resemblance to a circle. Nonetheless, here was the signpost that di-

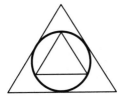

Figure 3-5. Circle with inscribed and circumscribed triangles

rected Archimedes to the most accurate determination of π in his time.

In *Measurement of a Circle*, Archimedes begins by imagining an equilateral triangle—one with equal sides—snugly inscribed within a circle, as depicted in figure 3-5. The triangle's three points lie on the circle itself, its three sides lie within the circle's interior. Nowadays, the triangle's overall perimeter is easy to compute: Using the basic relations of high school trigonometry, we first determine the length of any one side of the triangle; then, since all three sides are the same, we simply multiply this number by 3 to yield the perimeter. Archimedes does so (without the benefit of trigonometry, which had not yet been invented) and finds that the triangle's perimeter is 2.60 times as great as the diameter of the circle in which it is inscribed. Since the triangle's perimeter is here the proxy for the circle's circumference, the number 2.60 is Archimedes' initial estimate of the value of π. This, believe it or not, is progress. Admittedly, a triangle is not a circle, yet 2.60 at least sets a lower bound to the value of π. Like the ugly duckling who "became" a swan, Archimedes' seemingly clunky process will mature into something quite beautiful.

A corresponding upper bound to π also comes from a triangle, but one that neatly circumscribes the same circle. Such a figure has a computed perimeter of 5.19 times the circle's diameter.

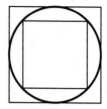

Figure 3-6. Circle with inscribed and circumscribed squares

Thus, Archimedes concludes that π lies somewhere between 2.60 and 5.19. Not too impressive, considering that π was already known at the time to be slightly more than 3.

Let's pick up Archimedes' process ourselves. Tighter limits on π come from a computation of the perimeters of inscribed and circumscribed squares, shown in figure 3-6. We start with the inscribed square. Drawing the diagonal that connects the opposite corners of the square divides it into two identical triangles, whose base is the diagonal and whose sides are two adjacent sides of the original square. Once again, the methods of trigonometry can be applied to either of these triangles, yielding the length of the square's side. Multiplying this length by 4 gives the square's perimeter. The entire process is repeated for the circumscribed square. (Archimedes would have slogged through a more involved procedure.) This time, we find that π is greater than the inscribed square's perimeter, 2.83, and less than the circumscribed square's perimeter, 4.00. Better than Archimedes' original estimate of π, yet still far short of what he would have considered acceptable.

Next draw inscribed and circumscribed pentagons, as in figure 3-7. Divide these likewise into triangular pieces and subject them to trigonometric analysis. This time the perimeters of the inscribed

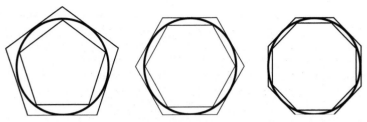

Figure 3-7. Circles with inscribed and circumscribed pentagons,
hexagons, and octagons

and circumscribed figures limit the value of π to between 2.94
and 3.64. Now we're truly closing in on π. Repeat the process
for inscribed and circumscribed hexagons, with six sides each,
also shown in figure 3-7; π falls between 3.00 and 3.45. Next,
octagons—eight-sided polygons; the bounds on π are now 3.06
and 3.31. Incredibly, Archimedes carries out this π-confining process up to a ninety-six-sided polygon. He doesn't, in fact, draw
such a multifaceted shape; instead he takes a theoretical shortcut
to render the shape's essential angle- and side-properties, and from
these, deduces its perimeter. Archimedes' final determination: π
must be greater than 3¹⁰⁄₇₁ and less than 3⅐; or in modern decimal
form, π lies somewhere between 3.1408 and 3.1429. His upper
bound, 3⅐, became the de facto standard for π into the Middle
Ages. Archimedes makes no claim to have fixed the value of π,
only to have established the narrow numerical range within
which it falls. Averaging his lower and upper bounds gives 3.1418,
deviating from the actual value of π by a mere 0.007 percent.

Archimedes applies a similar confinement technique, since
named the method of exhaustion, in his treatise *On the Sphere and
Cylinder* to determine the surface area of a sphere. (Euclid had already proven in book 12 of his *Elements* that the area of a circle is

rendered by the expression πR^2, where R represents the circle's radius.) Obviously, it's hard to measure a sphere's curved surface with a straight ruler or directly apply the theoretical metrics of plane geometry. So Archimedes invents a workaround. In a three-dimensional version of his polygon-as-circle substitution, he conjures a series of hollow, flat-sided figures to act as stand-ins for the sphere. To create such a figure, he mentally inscribes within the sphere a regular polygon with an even number of sides—a square, say. Then he rotates the square about one of its diagonals, such that the square's sides sweep out a three-dimensional figure. In the case of the square, the swept-out figure is a pair of cones joined at their bases. (Imagine holding between your fingers the opposite corners of a square card and twirling it.) The surface area of this inscribed double-cone, which Archimedes computes, provides a rough lower limit to the surface area of the surrounding sphere. Archimedes repeats the procedure for a hexagon, an octagon, and other even-sided shapes. The more sides the starting polygon has, the more interior space its swept-out, three-dimensional spin-off takes up and the more the new figure resembles the sphere in appearance. Each time, Archimedes computes the figure's surface area to arrive at a new lower bound on the sphere's surface area. As the series of inscribed figures grows in complexity, Archimedes converges on the surface area of the sphere itself. He concludes that the surface area of any sphere is uniquely linked to the sphere's radius: For a sphere of radius R, the surface area is given by the mathematical expression $4\pi R^2$ (written here in modern notation).

As a check, Archimedes conducts the analogous procedure with circumscribed polygons and reaches the same conclusion. Having done this, it is a straightforward, if time-consuming, matter to adjust the procedure to yield instead the sphere's volume:

$(\frac{4}{3})\pi R^3$. Archimedes must have felt supreme satisfaction that the results of his elegant investigation into the properties of spheres are two expressions of the utmost simplicity.

These examples give a flavor of Archimedes' analytical methods in geometry and, perhaps, a sense of why he is invariably named among the top mathematicians of all time. Granted, the swirling complexity of Archimedes' treatises is navigable by most who've studied plane geometry. Yet the arduousness of the journey is enough to scare away all but the most committed reader. An Archimedean treatise is the *Moby Dick* of mathematical writing; there is the compelling foreground narrative, virtuosic in its telling, but girded by an incredibly dense foundation of methodology—the mathematical equivalent of a Melvillean lesson in whale anatomy.

True appreciation of what Archimedes accomplished comes only from slogging through his lengthy proofs line by line. A simple statement about, say, the comparative volumes of a sphere and a cylinder requires an extensive dose of mathematical logic to prove by Euclid's strict standards. And reaching the end of the winding geometric yarn, one can only shake one's head in admiration and wonder: *How did he figure that out?* By what mathematical compass did Archimedes guide himself through this logical thicket when the route to the solution seems, at the outset, to be all but obscured? That Archimedes could divine such pathways time and time again must have seemed miraculous to his contemporaries, to Arabic mathematicians, and to the Renaissance thinkers who would rediscover his treatises centuries later.

Unlike his technological exploits, which practically beg for heroic telling, Archimedes' mathematical discoveries were too esoteric and complex to have seeped into the collective consciousness of ancient humanity. Only a relative handful of intellectuals at the time could appreciate his mathematical findings, much less comprehend them. Whereas Euclid's *Elements* immediately became the blockbuster text of geometry, Archimedes' works were relegated to the back shelf until the Greek mathematician Eutocius republished them with commentaries during the sixth century A.D., more than seven hundred years after Archimedes' demise. But by the mid-fourteenth century, with his treatises rendered in Latin, Archimedes was studied on a wider scale than ever before by scholars throughout Europe.

Consummate mathematician that he was, Archimedes was not above the occasional frolic across the game board of numbers and geometry. Among his surviving texts are several of a playful nature, seemingly designed more for amusement or mental challenge than for the advancement of mathematics. Their nominal purpose is to stretch the minds of royalty, Archimedes' overseas correspondents, and the numerate public. Yet these works, too, like his stock-in-trade Euclidean proofs, harbor a kernel of seriousness. Underneath their entertaining facades are probing meditations on important mathematical issues.

Chapter 4

NUMBER GAMES

Wherever there is number, there is beauty.

—Proclus, fifth-century Byzantine

philosopher–mathematician

A ROUND 2 I 6 B.C., shortly before his death, the septuage-
narian Archimedes completed a mathematical work for the
Syracusan monarch Gelon II, son of his longtime patron Hieron II.
In this curious tract, written for the layperson and titled *The Sand-
Reckoner,* Archimedes demonstrates his facility with the mathematics of large numbers—and, more to the point, expands the reach of
Greek arithmetic. To wit, Archimedes sets out to compute the
number of sand grains required to fill the universe. Yes, the universe. Along the way, he reveals his own pursuits in the field of
astronomy, including an account of his measurement of the Sun's
apparent diameter.

Archimedes' opening pitch in *The Sand-Reckoner* has a definite
ring of the carnival barker: "I will try to show you by means of
geometrical proofs, which you will be able to follow, that of the
numbers named by me . . . some exceed not only the number of

the mass of sand equal in magnitude to the Earth filled up in the way described, but also that of a mass equal in magnitude to the Universe." Sounds impressive, even today. But just how big is this universe that Archimedes intends to fill?

To make a lasting impression on his royal patron, Archimedes seeks a universe vaster than the commonly accepted geocentric system of his day, with its tightly nested heavenly spheres arrayed concentrically around a stationary Earth. So he selects a competing world system: the Sun-centered, or heliocentric, model of his elder contemporary, the astronomer-mathematician Aristarchus of Samos. Copernicus, the acclaimed inventor of the heliocentric model in the mid-1500s, was scooped by some eighteen hundred years when Aristarchus swapped the Earth and the Sun in the ancient Greek cosmic scheme. (To be fair, Copernicus developed his model independently of Aristarchus and endowed it with a mathematical rigor that far surpassed its classical forerunner.)

Nobody knows why Aristarchus advanced a central Sun when virtually all other Greeks held otherwise; his own treatise on the subject is lost. Presumably, his reasons were similar to those of Copernicus: The Sun is both larger than the Earth and unique among the celestial bodies the Greeks called *planets* in its brilliance and warmth. The Sun, as Copernicus saw it, is the "lantern" that illuminates the heavens and, logically, sits on the cosmic throne. Were it not for Archimedes' mention in *The Sand-Reckoner*, we would know even less than the little we do about Aristarchus's heliocentric theory. Archimedes has nothing to say about the relative merits of the geocentric model versus its heliocentric counterpart. He is interested in only one feature of Aristarchus's heliocentric model: its vastness.

First let's lay the groundwork for what Aristarchus has to say

about the extent of the universe Archimedes proposes to fill, for
he approaches the issue in an oblique and somewhat problematic
fashion. In general, how can one effectively convey the size of an
object? One method is to state its dimensions outright: The house
is thirty feet wide. Another method is to compare its size to that
of another object or, equivalently, to form the size *ratio* between
the objects: The shopping mall is twice the length of a football
field; or the ratio between the length of the shopping mall and the
length of the football field is 2:1. Yet a third method is to express
one size ratio in terms of another: The tree's height is to the man's
height as the man's height is to the child's height. In this case,
if the man is six feet tall and the child three feet tall, the ratio of
their heights is 2:1; since the tree's height bears this same ratio to
the man's height, the tree is twelve feet tall. It is the last of these
methods that Aristarchus chooses to reveal the size of his universe.
Here, in modernized form, is what Archimedes claims Aristarchus
wrote.

> The distance of the stars bears the same relation to the di-
> ameter of the Earth's orbit as the surface of a sphere bears to
> its center.

To the modern ear, Aristarchus's statement is reminiscent of
one of those thorny problems on the SAT. Parsing it, we see that
the statement contains two ratios: The first ratio is a purely physical one, between the *distance of the stars* and the *diameter of the
Earth's orbit*; the second ratio is a purely mathematical one, between the *surface of a sphere* and the *center of a sphere*. These two ratios, Aristarchus claims, are equal, which we can cast as a simple
equation.

distance of the stars : diameter of the Earth's orbit =
surface of a sphere : center of a sphere

Compute the second ratio, the mathematical one, and you simultaneously have learned the first ratio, the physical one. Here Aristarchus apparently offers Archimedes a way to compute the extent of the heliocentric universe—the equation's *distance of the stars*—in terms of the diameter of the Earth's orbit. Specifically, how many times bigger is the universe compared to the Earth's orbit? We can answer that question by computing the second ratio. Or can we?

The geometrically savvy reader will already have recognized that the second ratio, between the *surface of a sphere* and the *center of a sphere*, is patently absurd. Yet similar phrasing was often used by Greek astronomers when expressing the immensity of the heavens. (By *surface of a sphere*, we'll take Aristarchus's meaning to be the sphere's diameter.) The center of a sphere is a point, which geometrically speaking has no size at all. Thus a sphere's diameter is *infinitely* larger than its center, because any number divided by zero yields infinity. In other words, the second ratio turns out to be infinite. Taken literally, Aristarchus's statement implies that the stars are infinitely far away. Perhaps that is his intended meaning, perhaps not.

An infinite universe does not suit Archimedes, who at this point might have identified with the hapless protagonist in the physicist George Gamow's limerick.

There was a young fellow from Trinity
Who took [the square root of infinity]

But the number of digits
Gave him the fidgets;
He dropped Math and took up Divinity.

In *The Sand-Reckoner,* Archimedes wishes to perform a calculation that involves the purported extent of the universe. He is
unable to deal with one that is infinitely large, so he fudges
his assumptions a bit. He chooses a less literal interpretation of
Aristarchus's enigmatic phrase comparing the surface to the center of a sphere. According to Archimedes, Aristarchus must have
meant something like this:

> The distance of the stars bears the same relation to the diameter
> of the Earth's orbit as the diameter of the Earth's orbit bears to
> the diameter of the Earth.

Or, in equation form:

$$\text{distance of the stars : diameter of the Earth's orbit} =$$
$$\text{diameter of the Earth's orbit : diameter of the Earth}$$

Voila! The second ratio is no longer infinite, but a straightforward comparison between two measurable quantities: the *diameter
of the Earth's orbit* and the *diameter of the Earth.* These were numbers for which crude estimates already existed (from Aristotle,
among others). Using generous assumptions, Archimedes reasons
that the diameter of the Earth's orbit does not exceed 10,000
Earth-diameters. (In reality, he is still too low by a factor of two.)
The previous equation becomes:

distance of the stars : 10,000 Earth-diameters =
10,000 Earth-diameters : 1 Earth-diameter

Therefore, according to Archimedes, the stars on the celestial sphere are 100 million Earth-diameters away, which dwarfs the estimate of 10,000 Earth-diameters proposed by the geocentrists. (Yet it still falls far short of the true distance to even the nearest star: over 3 *billion* Earth-diameters.) Next Archimedes converts his cosmic distance to *stadia*, a terrestrial measure used by the ancients and roughly equivalent to a tenth of a mile. For this calculation, he again fudges his assumptions. Remember, Archimedes was trying to impress King Gelon by filling the biggest possible universe, so he unabashedly gooses up the numbers for maximum effect. He takes the Earth's diameter to be nearly 1 million stadia, which he freely admits is tenfold larger than the commonly accepted value at the time. At last, Archimedes deduces that the Aristarchian universe is about 100 trillion stadia in radius, roughly 10 trillion miles.

This is only the halfway point of the problem! Archimedes has to count how many sand grains would be required to fill such an expanse. That, of course, depends on the minuteness of the grains. Archimedes had no microscope with which to view samples of sand, much less one fitted with a measuring stage. He approaches the issue by comparing sand to a mote he could more easily measure: a poppy seed. A line of twenty-five poppy seeds, he informs King Gelon, matches the width of his forefinger; thus, on average, a poppy seed is 1/25 of a finger-width across. The volume of one such seed, he speculates, would hold up to ten thousand sand grains—a myriad, in the Greek number system. (Evidently, Syracusan sand is finer than the best the Caribbean has to offer.) Yet even such a tiny speck is too bloated for Archimedes' purpose; his goal is to dazzle Gelon at the end of this numerate tale with the largest imaginable number of sand grains that would fill the universe. The tinier a poppy seed, the tinier the diminutive sand grain, and the more grains it would take to pack the cosmos. Therefore, Archimedes jettisons his measured poppy-seed diameter and claims instead that a seed spans a mere 1/40 of a finger-width. A sand grain is now commensurably smaller; it takes some four times as many of them to fill a given space than it did before.

Having sized up both the vast universe and the minuscule sand grain, Archimedes arrives at his conclusion: To fill space all the way out to the sphere of the stars requires one-thousand trillion trillion trillion trillion trillion sand grains. In modern powers-of-ten notation, the requisite number of sand grains would be written 10^{63}, that is, the digit 1 followed by sixty-three zeroes. But how to inform King Gelon of this colossal answer? The decimal number system was unknown to the ancient Greeks. The number-naming capacity of their system, which expresses numbers as strings of letters, petered out at numbers above 100 million, or 10^8. It was virtually incapable of depicting a number as large as 10^{63}. Archimedes saw this not as an obstacle but as an opportunity. He would augment the standard counting system with the capability to name a number larger than any that had been named before.

Recall that the ancient Greek counting system assigns letter equivalents to specific numbers: 1, 2 , 3, and so on, up to 9 are, respectively, α (alpha), β (beta), γ (gamma), and succeeding letters up to θ (theta); 10, 20, 30, and so on, up to 90 are ι (iota), κ (kappa), λ (lambda), and succeeding letters up to ϙ (the archaic koppa); 100, 200, 300, and so on, up to 900 are ρ (rho), σ (sigma), τ (tau), and succeeding letters up to ϡ (the archaic letter san). Numbers between 1,000 and 9,000 are expressed by recycling the first nine letters, but with a leading subscripted vertical line: 1,000 is $_|\alpha$; 2,000 is $_|\beta$; and so on. The number 10,000, called a myriad, is designated by the symbol M. Thus, the number 12,333 is written $M_|\beta\tau\lambda\gamma$. There are various alphabetic codes written above the M to designate numbers higher than 10,000, all the way up to a myriad-myriads, or 100 million (10^8).

Archimedes defines numbers up to 10^8, or a myriad-myriads, as those of the *first order*. Then, using 10^8 as the new starting unit, he defines numbers belonging to the *second order*: 10^8, 2×10^8 (that is, 200 million), 3×10^8 (300 million), and so on, up to $10^8 \times 10^8$, or 10^{16}. Archimedes calls this last number a "myriad-myriads of the unit of the first order." Numbers of the *third order* use 10^{16} as the starting unit: 10^{16}, 2×10^{16}, 3×10^{16}, and so on up to $10^8 \times 10^{16}$, or 10^{24}. This last is called a "myriad-myriads of the unit of the second order." Not to skimp on the range of number-names, Archimedes explains how to extend his system far beyond the requisite 10^{63} sand grains, up to the dizzying numerical heights of $10^8 \times 10^8 \times \cdots \times 10^8$, where the ellipsis substitutes for 99,999,997 additional multiplications of 10^8, abbreviated $(10^8)^{10^8}$. Archimedes terms this number a "myriad-myriads of the unit of the myriad-myriadth order."

At this point, King Gelon's head must have been spinning. Yet, in characteristic fashion, Archimedes plows ahead. He defines his last-named number and all the numbers up to it as but the *first period* of a vastly larger counting scheme! Each successive period covers more numerical ground than its predecessor. It is not until Archimedes has tallied a myriad-myriad periods that his numerical cornucopia has exhausted itself. The last and largest Archimedean number is "a myriad-myriad units of the myriad-myriadth order of the myriad-myriadth period," or $((10^8)^{10^8})^{10^8}$. Archimedes concludes *The Sand-Reckoner* by telling King Gelon that "these things will appear incredible to the numerous persons who have not studied mathematics; but to those who are conversant therewith and have given thought to the distances and the sizes of the earth, the sun, and the moon, and of the whole of the cosmos, the proof will carry conviction. It is for this reason that I thought it would not displease you either to consider these things." Whether Gelon went to sleep that night inspired or with a headache, history does not report.

The infamous Archimedean puzzler known as the Cattle Problem resurfaced in 1773, after the German critic and dramatist Gotthold Ephraim Lessing found its text in a manuscript at the Herzog August Library in Wolfenbüttel, Germany, where he worked as librarian. "A problem," it begins matter-of-factly, "which Archimedes devised in epigrams, and which he communicated to students of such matters at Alexandria in a letter to Eratosthenes of Cyrene." The Cattle Problem was well known in the ancient world; Cicero refers to it twice as a metaphor for complexity. Some historians have suggested that Archimedes created the problem in a rush of competitive adrenaline after Apollonius produced a treatise on the naming of large numbers, Archimedes' preeminent domain.

The cattle in question were the property of the Sun-god, Helios, and make their appearance in book 12 of the *Odyssey*. The immense herd is grazing in a Sicilian pasture when Odysseus's starving crew arrives and foolishly slaughters several of the cattle for food. With a lightning bolt, Zeus exacts his revenge on the hapless mortals. And, with a mathematical bolt of his own, Archimedes exacts his revenge on those who dare match wits with him.

Here is the first part of the Cattle Problem, in the form Gotthold Lessing found it.

If thou art diligent and wise, O stranger, compute the number of cattle of the Sun, who once upon a time grazed on the fields of the Thrinacian [three-cornered] isle of Sicily, divided into four herds of different colors, one milk white, another a glossy black, a third yellow and the last dappled. In each herd were bulls, mighty in number according to these proportions: Understand, stranger, that the white bulls were equal to a half and a third of the black together with the whole of the yellow, while the black were equal to the fourth part of the dappled and a fifth, together with, once more, the whole of the yellow. Observe further that the remaining bulls, the dappled, were equal to a sixth part of the white and a seventh, together with all of the yellow. These were the proportions of the cows: The white were precisely equal to the third part and a fourth of the whole herd of the black; while the black were equal to the fourth part once more of the dappled and with it a fifth part, when all, including the bulls, went to pasture together. Now the dappled in four parts were equal in number to a fifth part and a sixth of the yellow herd. Finally the yellow were in number equal to a sixth part and a seventh of the white herd. If thou canst accurately tell, O stranger, the number of cattle of the Sun, giving separately the number of well-fed bulls and again the number of females according to each color, thou wouldst not be called unskilled or ignorant of numbers, but not yet shalt thou be numbered among the wise.

In other words, given this long-winded list of clues, compute the overall number of cattle and the number of males and the number of females of each color. In Archimedes' opinion, jumping this high hurdle merits but a laggard's trophy labeled NOT

UNSKILLED OR IGNORANT OF NUMBERS. In fact, as worded, the first part of the problem has an infinite number of solutions. For those wishing to gain entry into the ranks of the wise, Archimedes raises the bar to Olympian height; he adds further criteria to reduce the array of possible solutions down to a single, unique set of numbers. Here is the second part of the problem.

But come, understand also all these conditions regarding the cattle of the Sun. When the white bulls mingled their number with the black, they stood firm, equal in depth and breadth, and the plains of Thrinacia, stretching far in all ways, were filled with their multitude. Again, when the yellow and the dappled bulls were gathered into one herd they stood in such a manner that their number, beginning from one, grew slowly greater till it completed a triangular figure, there being no bulls of other colors in their midst nor none of them lacking. If thou art able, O stranger, to find out all these things and gather them together in your mind, giving all the relations, thou shalt depart crowned with glory and knowing that thou hast been adjudged perfect in this species of wisdom.

The first part of the problem is tackled by translating the expressed proportions of the various cattle into a set of equations.

- "white bulls were equal to a half and a third of the black [bulls] together with the whole of the yellow [bulls]" becomes the equation
  $$WB = (\tfrac{1}{2} + \tfrac{1}{3})\, BB + YB$$

- "black [bulls] were equal to the fourth part of the dappled [bulls] and a fifth, together with, once more, the
  whole of the yellow [bulls]" becomes the equation
  $$BB = (\tfrac{1}{4} + \tfrac{1}{5}) \, DB + YB$$
- "the remaining bulls, the dappled, were equal to a sixth
  part of the white [bulls] and a seventh, together with all
  of the yellow [bulls]" becomes the equation
  $$DB = (\tfrac{1}{6} + \tfrac{1}{7}) \, WB + YB$$
- "The white [cows] were precisely equal to the third part
  and a fourth of the whole herd of the black" becomes
  the equation
  $$WC = (\tfrac{1}{3} + \tfrac{1}{4}) \, (BB + BC)$$
- "the black [cows] were equal to the fourth part once
  more of the dappled and with it a fifth part, when all,
  including the bulls, went to pasture together" becomes
  the equation
  $$BC = (\tfrac{1}{4} + \tfrac{1}{5}) \, (DB + DC)$$
- "the dappled [cows] in four parts [i.e., all of them] were
  equal in number to a fifth part and a sixth of the yellow
  herd" becomes the equation
  $$DC = (\tfrac{1}{5} + \tfrac{1}{6}) \, (YB + YC)$$
- "the yellow [cows] were in number equal to a sixth part
  and a seventh of the white herd" becomes the equation
  $$YC = (\tfrac{1}{6} + \tfrac{1}{7}) \, (WB + WC)$$

By standard methods of algebra, these equations generate the
following expressions for the number of cattle of various types.

white bulls $WB = 10,366,482 \, k$
black bulls $BB = 7,460,514 \, k$

yellow bulls	$YB = 4{,}149{,}387\ k$
dappled bulls	$DB = 7{,}358{,}060\ k$
white cows	$WC = 7{,}206{,}360\ k$
black cows	$BC = 4{,}893{,}246\ k$
yellow cows	$YC = 5{,}439{,}213\ k$
dappled cows	$DC = 3{,}515{,}820\ k$

Here the letter k represents an integer that multiplies its preceding numerical coefficient. The smallest solution that satisfies the first part of the Cattle Problem occurs when k equals 1, although any integer greater than 1 is just as valid a solution. If k equals 1, the total number of cattle adds up to 50,389,082. However, the second part of the problem imposes an additional pair of constraints on k that boosts its value into the stratosphere.

The first of these constraints is expressed by the phrase, "When the white bulls mingled their number with the black, they stood firm, equal in depth and breadth." Properly parsed, this means that the combined herd of white and black bulls, when uniformly distributed among rows and columns, covers an area whose shape is a square. Or mathematically, the number of white bulls WB (or 10,366,482 k, from the solution to the first part of the problem) added to the number of black bulls BB (7,460,514 k) yields a number that is a square number, that is, an integer formed by multiplying another integer by itself. Examples of square numbers are 9, which is 3×3, and 100, which is 10×10. Thus, the integer k must be chosen so that the sum 10,366,482 k + 7,460,514 k, or 17,826,996 k, is a square number.

Before proceeding, let's consider a simpler example, say, the statement that 10 k is a square number. Obviously, the condition is fulfilled when $k = 10$, because $10 \times 10 = 100$ results in a square number.

But a square number likewise appears when our selected k value, 10, is multiplied by a second number that is itself a square number, like 4 or 9 or 25 or, in general, any integer of the form $m \times m = m^2$. Thus, to make our original number 10 k into a square number, k must have a value of 10 m^2, where m is any integer.

Now back to the problem at hand: 17,826,996 k must be a square number. By analogy with the previous example, we would expect that to produce the intended result, k should have a value of 17,826,996 m^2. But if we first factor the coefficient 17,826,996—that is, find the series of prime integers that, when multiplied together, produce the number 17,826,996—we find that 17,826,996 has a square number hidden within it. We can excise that square from the coefficient and toss it into the m^2 term where it properly belongs. The prime factors of 17,826,996 are $2 \times 2 \times 3 \times 11 \times 29 \times 4657$. There's the square number: 2×2. Let's remove it from the number and incorporate it in the m^2 term. The remaining factors $3 \times 11 \times 29 \times 4657$ multiply out to 4,456,749. Thus recast, this first constraint on the value of k produces the equation:

$$k = 4{,}456{,}749\, m^2, \text{ where } m \text{ is a positive integer}$$

The second Archimedean constraint on k arises in the phrase "when the yellow and the dappled bulls were gathered into one herd they stood in such a manner that their number, beginning from one, grew slowly greater till it completed a triangular figure." This means that the number of yellow bulls YB (4,149,387 k, again from the first part of the problem) added to the number of dappled bulls DB (7,358,060 k) yields a triangular number—one formed by adding successive integers up to a given limit:

$1 + 2 + 3 + 4 + \cdots + n$. This time we must ascertain the value of k that turns the sum $4{,}149{,}387\ k + 7{,}358{,}060\ k$, or $11{,}507{,}447\ k$, into such a triangular number.

A useful algebraic equivalence is that, for identical values of n, the sum $1 + 2 + 3 + 4 + \cdots + n$ produces the same result as does the expression $n(n+1)/2$. (Try an example; it works.) This means that

$$11{,}507{,}447\ k = n(n+1)/2$$

Substituting the previous constraint condition on k, we obtain

$$11{,}507{,}447 \times 4{,}456{,}749\ m^2 = n(n+1)/2$$

Completing the multiplication on the left side leaves us with the final constraint expression

$$102{,}571{,}605{,}819{,}606\ m^2 = n(n+1)/2$$

The goal is threefold: find the values of m and n that satisfy this equation; substitute that value of m into the previous constraint equation to compute k; substitute that value of k into the cattle equations from the problem's first part to determine (finally!) the number of cattle of each kind.

You're in good company if you can't make headway with the Cattle Problem. It's likely that Archimedes himself couldn't solve it, although he was probably convinced that a solution was possible. More than a hundred years elapsed after Lessing's resurrection of the Cattle Problem before anyone came up with even a partial solution. In 1880, the German mathematician A. Amthor deduced that the total number of cattle is a 206,545-digit integer that begins with the digits 776—surely not the degree of precision Archimedes had in mind. (Archimedes didn't delve into the practical dimensions of his problem; one commentator pointed out that even if the cattle were each shrunk to the size of a bacterium, the herd would not fit within the space of a sphere 10,000 light-years across.)

Fifteen years after Amthor, in the *American Mathematical Monthly*, A. H. Bell specified the first 31 and the last 12 digits of the overall number of cattle: 7,760,271,406,486,818,269,530,232,833,209 . . . 719,455,081,800, where the ellipsis stands in for 206,502 missing digits. (The digits 09 before the ellipsis have since been revised to 13.) Bell and his two compatriots, Edmund Fish and George H. Richards, the sole members of the Hillsboro, Illinois, Mathematical Club, had labored four years to develop their answer. They duly note that their answer, if fully printed out, would extend half a mile.

There the Cattle Problem stood until the advent of the computer. In 1965, mathematicians at the University of Waterloo, in Ontario, Canada, generated a complete solution—all 206,545 digits—after their IBM Model 7040 and 1620 computers churned away for almost eight hours. The printed solution takes up forty-two large-format computer sheets, which are stored in a box within the Archives of American Mathematics at the University of Texas at Austin. In 1981, the ancient Cattle Problem was considered a sufficiently rigorous test for the new Cray 1 supercomputer at the Lawrence Livermore Laboratory. The fastest computer in the world took a full ten minutes to solve the puzzle.

Among a series of Archimedean treatises discovered during the twentieth century is a crumbling fragment of a work called *Stomachion*. Although the word stems from the Greek for stomach, the object of the treatise is neither anatomy nor physiology, but what may be the world's oldest tiling puzzle. Although it was popularly known during antiquity as *loculus Archimedius*—Archimedes' box— Archimedes might not have been the puzzle's inventor but was merely drawn to its geometric intricacies.

The puzzle resembles a tangram game. In its ancient incarnation, depicted in figure 4-1, a square of ivory is sliced in a prescribed way to produce a total of fourteen three-sided, four-sided, or five-sided polygons. (In 1926, the *New York Times* printed instructions on how to cut the Stomachion from cardboard.) These polygons can be rearranged in a variety of ways to reconstitute the original square. Or they can be combined in seemingly limitless fashion to form shapes of animals, people, or everyday objects. The fourth-century Roman poet and statesman Ausonius writes that he created an elephant, boar, flying goose, armed gladiator, squatting huntsman, barking dog, tower, and tankard.

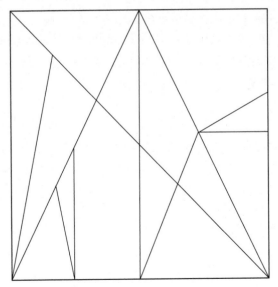

Figure 4-1. Archimedes' Stomachion tiling puzzle

As laid out by Archimedes, the Stomachion's polygonal tiles also bear a curious mathematical relationship to the initial square from which they are cut. Imagine subdividing the square into a 12-by-12 grid. The square's area is the product of its length and width, in this case, 12 × 12, or 144. Remarkably, the area of each polygonal tile is an integer number: Within the square's total area of 144 are two polygons of area 3; four of area 6; five of area 12; plus one each of areas 9, 21, and 24.

In his treatise—or at least in the extant fragment—Archimedes tries to ascertain the multitude of ways the Stomachion polygons fit together to form a square, a branch of mathematics now called combinatorics. In 2003, the barbecue grill salesman and game hobbyist Joe Marasco offered a one-hundred-dollar prize to anyone who identified all the possible solutions of the puzzle. Before year's end, the veteran puzzle maker Bill Cutler used a computer program he devised to determine that there are 536 unique ways to form the Stomachion square from its polygonal pieces.

The Sand-Reckoner, the Cattle Problem, and *Stomachion* exemplify Archimedes' "off-hours" pursuits, yet none of them is trivial by any measure. Coupled with his more formal treatises, they comprise an impressive body of work that ranges widely over what was then the fledgling realm of mathematics. But antiquity's Master of Thought reached far beyond the abstract complexities of shapes and numbers. He delved deeply into the real-world realms of physics and engineering and emerged with foundational results that still astonish today.

Chapter 5

EUREKA MAN

How can it be that mathematics, being after all a product of human thought independent of experience, is so admirably adapted to the objects of reality?
—ALBERT EINSTEIN, *SIDELIGHTS ON RELATIVITY*

ARCHIMEDES' FIRST LOVE was pure mathematics, if the historian Plutarch is to be believed. Yet his reputation, during both antiquity and the Middle Ages, rested at least initially on his resourcefulness as a practical physicist and engineer. With his investigations of the properties of the lever and of centers of gravity, Archimedes originated the quantitative study of statics, the branch of engineering that assays the steadiness of bridges and buildings. Likewise, his treatises on buoyancy and the stability of floating objects gave birth to the realm of physics called hydrostatics.

The basics of hydrostatics are set down in book 1 of Archimedes' treatise *On Floating Bodies*. Book 2 presents a mathematical tour de force on the tipping tendencies of floating objects, in which Archimedes deduces to what extent an object can list before it turns topsy-turvy into the water. Of his practical works, book 2 rises farthest above the level of mathematical sophistication and physical ingenuity that prevailed during the classical age. Yet despite the topic, this was no practitioner's guide to ship design, for none of the bobbing objects studied by Archimedes remotely resembles a real vessel: lopped-off spheres and variously shaped paraboloids (think of rounded ice cream cones—or better yet, the Coneheads on *Saturday Night Live*).

The flotation of ships, sharks, and swimmers is governed by the tug-of-war between gravity, which tends to pull an object down, and buoyancy, which tends to push it up. The force of gravity is exerted between an object and the Earth's mass—it's what we call the object's *weight*—and it behaves no differently in water than in air or in a vacuum. The countering buoyancy force is a consequence of the increase in fluid pressure with depth. If you've ever dived to the bottom of a pool, you feel this fluid pressure on your eardrums. Buoyancy arises because the pressure force pushing up against the bottom of a submerged object is greater than that pushing down against the top. The pressure forces that act against the object's sides cancel each other out and don't affect the object's buoyancy. The net result is that buoyancy tries to float an object against gravity's tendency to sink it.

Archimedes tells us that a submerged object weighs less than it does in air. The reason for this difference is the action of the buoyancy force, which offsets a portion of gravity's downward pull. It's as though you were lifting the object slightly with your finger while weighing it. In fact, the numerical difference between the object's weight in air and its submerged weight is an accurate measure of the magnitude of the buoyancy force.

Whether an immersed object floats or sinks depends on the

magnitudes of the opposing forces of gravity and buoyancy. If the weight of the immersed object—that is, gravity's downward pull on it—is greater than the upward buoyancy force, the object will sink; if the weight of the immersed object is less than the buoyancy force, the object will float. That's why a stone sinks but an ice cube floats; for the same volume, the stone packs more weight than the ice cube. A completely submerged object whose weight and buoyancy are equal can hover at any level within the water, like a fish. A submarine can dive or rise at will, by either filling its ballast tanks with seawater (making itself heavier) or pumping the seawater out (making itself lighter).

Archimedes' contribution to this purely descriptive tale of competing forces was to extract from it a mathematical kernel that reveals the magnitude of the buoyancy force. His insight is expressed in a statement since named Archimedes' Principle: *An immersed object is buoyed by a force equal to the weight of the fluid it displaces.* With this, the contest of forces governing the flotation or sinking of an object is completely determined; the object's weight can simply be measured on a scale, while the buoyant force on it can be linked to the object's volume. The larger that volume, the more buoyancy the object will experience. Don't be distressed if this all seems rather obtuse. Why the floating ability of a ship has anything to do with the amount of water its hull pushes aside is not obvious and requires explanation.

Every ship rides at a certain level in the water, depending on its hull shape and how heavy it is. A heavily laden ship rides deeper in the water than when empty. It sinks to the point that its hull displaces an amount of water whose weight equals that of the ship, that is, until gravity's force pulling the ship down and water's buoyancy pushing it up are in balance. A one-thousand-ton ship, for example, will sink into the sea until its hull displaces a volume of water that weighs one thousand tons; at that point, the upward buoyant force is sufficient to offset the downward pull of gravity. A ship twice as heavy will ride at a level such that its hull displaces two thousand tons of water.

In fact, setting a ship into the sea means that the water formerly occupying the ship's immersed volume must go elsewhere. The water is literally displaced. Where does it go? It is pushed into its surroundings. But water is incompressible; the displaced water itself displaces more water around it, and that water in turn displaces other masses of water. If we consider the sea as a vast, yet finite, basin, ultimately the displaced water has no place to go but up. (Think of the rise in water level as Archimedes settles into his bathtub.) Thus, every time a ship is lowered into the ocean, sea level rises imperceptibly against the weight of the atmosphere. So it's really the push-back of all this air on the Earth's bodies of water that causes ships to float. Insubstantial as it may seem, air does have weight, especially when you consider that we live at the bottom of a veritable ocean of air—our atmosphere. At sea level, the atmosphere presses on every square inch of your body—and the sea's surface—with a weight of 14.7 pounds. When it comes to flotation, a ship is no different than an iceberg in the sea or an ice cube in a drink. Scientists have discovered that continents, too, behave in strict accordance with Archimedes' Principle; they ride, partly immersed, upon Earth's semimolten mantle layer like outsize, rocky barges.

Archimedes' hydrostatic researches are intimately linked with the popular *Eureka* story involving the counterfeit crown. As told by the first-century B.C. Roman architect Vitruvius, the colorful tale of Archimedes' insight is almost certainly false, both in its technical details and in the notorious depiction of the great sage running naked through the streets. Probably true is King Hieron's directive to Archimedes: determine whether the goldsmith had replaced, with an equal weight of lesser-grade metal, some of the gold he had been given for a wreathlike crown—and proceed without sampling or otherwise defacing the crown.

Gold is heavy stuff: A three-inch gold cube weighs in at nearly nineteen pounds. (Talk about the burdens of high office; no wonder George Washington refused to wear royal trappings.) Archimedes knew that gold is heavier than other metals like silver or copper, which the artisan might have snuck into his raw material. Vitruvius's version of events relates how Archimedes needed only to measure the *volume* of the suspect crown. If the crown took up the same space as an equivalent weight of pure gold, then it was made exclusively of gold; but if it took up more space than the equivalent weight of gold, it was formed from an alloy.

What purportedly set Archimedes barreling through ancient Syracuse in unclothed euphoria was his bath-time realization that the crown's volume might be gauged by immersing it in a vessel full of water and measuring the overflow. The trouble with Vitruvius's scenario is twofold. First, Archimedes' epiphany about volume measurement seems rather mundane compared to his other *Eureka*-worthy insights; it's hard to imagine him exulting in the streets about it. Second, as Galileo pointed out in 1586, to prove such metallurgical fakery in the manner Vitruvius ascribes to Archimedes would have required a degree of measurement precision unavailable in ancient times. The difference in the volume of overflow produced by immersing a pure gold crown versus an adulterated one would have been too minute to discern.

Galileo could not believe that the vaunted Archimedes would have posed such a flawed technique. "This seems, so to say, a crude thing," mused the twenty-two-year-old Galileo, "far from scientific precision; and it will seem even more so to those who have read and understood the very subtle inventions of this divine man in his own writings; from which one most clearly realizes how inferior all other minds are to Archimedes'." The overflow method, Galileo concluded, was a misstatement by Vitruvius, who was evidently insensitive to the practicalities of the problem.

More likely, as Galileo proposed in his first-ever scientific pamphlet, *La Bilancetta (The Little Balance)*, Archimedes used a hydrostatic version of a beam balance, a device that resembles the proverbial scales of justice. From one arm of the beam, Archimedes suspended the suspect crown, and from the other arm an equivalent weight of pure gold. The two objects being of equal weight and situated equal distances from the central pivot, the beam would have been stable in any orientation. A nudge from Archimedes' finger would have brought it to its proper starting orientation—perfectly horizontal.

At this point, Archimedes might have immersed the suspended objects into vessels of water. There were two possible outcomes. If the crown was indeed made of pure gold, the buoyant forces on it and on the matching gold slug would be identical and the balance would remain horizontal. (Why? The same weight of the same substance must occupy the same volume, whatever its shape; thus, according to Archimedes' Principle, both the crown and the gold slug would be buoyed identically by the water.) The other possible outcome is for an adulterated crown. Such a crown would have had a slightly larger volume than one of pure gold because silver or copper takes up more space than an equivalent weight of gold. Immersed in water, a larger-volume crown would be buoyed more strongly than the matching gold slug; the beam of the balance would tip, the crown side higher than the slug side. Evidently, the balance did tip, indicating that the artisan had replaced some of the gold he had been given with a less dense metal.

In his paper, Galileo lays out the presumed design of Archimedes' hydrostatic balance, down to wrapping the arms of the apparatus with thin wire to create a hyperprecise measurement scale. "Since the wires are very fine, as is needed for precision, it is not possible to count visually, because the eye is dazzled by such small spaces. To count them easily, therefore, take a most sharp stiletto and pass it slowly over said wires. Thus, partly through our hearing, partly through our hands feeling an obstacle at each turn of wire, we shall easily count said turns." Galileo doesn't tell us whether he put his envisioned Archimedean balance to the test. But a modern analysis reveals that such a device applied to Hieron's crown could reasonably have detected as little as 6.5 percent silver intermixed with gold. If Galileo was right, the larcenous goldsmith stood little chance against the detecting skill of Archimedes.

Another Archimedean endeavor that has entered the realm of myth is his work with levers and pulleys. Archimedes' bold claim that he could move the Earth if only given a place to stand resonated with admirers like Galileo, whose own groundbreaking explorations gave him a similar god's-eye outlook on nature. Archimedes' proof of the law of the lever, in his treatise *On the Equilibrium of Planes*, remains one of the cornerstones of his reputation. Archimedes didn't invent the lever. That occurred in the dim recesses of human history from what might have been the chance confluence of the need to move a boulder and the availability of a felled log. Indeed, by Archimedes' time, the practicalities of the lever were familiar to anyone engaged in construction. It was common knowledge, for instance, that to balance two weights on a lever beam (essentially, a seesaw), the heavier weight had to be placed closer to the fulcrum than the lighter one. Yet the numerical underpinnings of the lever's behavior had not been fully explored.

When Archimedes cast his mathematical net over this narrow branch of mechanics, he characteristically hauled in a richer catch than predecessors who had plumbed the same depths. His treatment of the lever is stunningly innovative for its time and came to influence the methods of pre-Renaissance and Renaissance mathematicians.

Archimedes simplifies a physical problem through abstraction, dispensing with all material properties not crucial to the problem's solution. An object sheds its defining characteristics—shape, size, composition, texture, appearance—save one: its weight, or what Archimedes, in proper mathematical fashion, calls its *magnitude.* Here we encounter weight in its abstract sense, weight as a pure number. Instead of filling up the form of the object, the Archimedean weight is condensed into a geometric point, which physicists now term the object's center of gravity. The center of gravity represents a mathematically defined average of the distribution of matter within an object. Suspended from this point, an object would remain motionless in any orientation. This is easier to picture for flat objects than for solid objects, where a narrow hole would have to be drilled to accommodate a suspension thread.

A symmetrical object whose weight is evenly distributed—a uniform disk, say—typically has a center of gravity at its geometric center. The center of gravity of an object with a lopsided distribution of weight will be offset toward the object's heavier side:

A sledge hammer balances at a point closer to its head than to the end of its handle. Some objects, such as a crescent, chevron, or torus, might have a center of gravity somewhere in the space outside their material borders. The center of gravity can apply to a system of objects considered as a whole: A fly landing near the edge of a disk will shift the center of gravity of the fly-disk system from where it was for the disk alone. In his treatise *On the Equilibrium of Planes*, Archimedes deduces the center of gravity for an array of different forms, including triangles, parallelograms, and parabolic sections.

Archimedes abstracts the lever one step further: The lever beam is no longer a physical plank but a uniform line, a mere holder of weights that plays no active role in the mathematical disposition of the device. And the fulcrum? Again, an abstraction, a phantom-spike atop which the impalpable lever beam pivots. Archimedes' world of levers is rife with such insubstantial entities that arrange and rearrange themselves to the mathematician's will. Here, the creak and groan of real levers and the sweat of muscular effort are banished.

From experience, Archimedes envisioned the law of the lever: *To balance a pair of unequal weights, the weights must be placed at distances from the fulcrum that are in inverse proportion to their magnitudes.* In mathematical terms, if D represents the distance from the fulcrum to the heavier weight W, and d represents the distance from the fulcrum to the lighter weight w, the lever will come into balance, or equilibrium, when the ratio D/d equals the ratio w/W, as depicted in figure 5-1. For example, if the heavier object weighs three times as much as the lighter one, as in figure 5-2, then a lever will balance if the heavier object is one-third as far from the fulcrum as the lighter object.

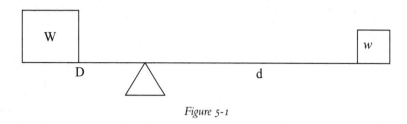

Figure 5-1

To prove the law of the lever, Archimedes plays the geometric craftsman. He sizes up the task, intuits a stepwise plan for its completion, and gathers the necessary deductive tools. He appeals to the inexorable logic of symmetry. Through his eyes, we see an asymmetric lever bearing unequal weights transform itself into a symmetric lever bearing equal weights. Archimedes stuffs a handkerchief into his top hat and, with a tap of the mathematical wand, pulls out a rabbit. Observe how he does it.

Archimedes starts off by stating the obvious: *Equal weights situated equal distances from the fulcrum are in equilibrium.* No argument there. From this and a couple of other straightforward assumptions, he deduces an intermediate proposition: *The center of gravity of two equal weights lies at the midpoint of the line connecting their individual centers of gravity.* But this proposition, which holds for two equal weights, remains true when those weights are joined by another pair of equal weights added symmetrically on both

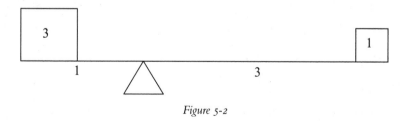

Figure 5-2

sides of the fulcrum. A seesaw is balanced if you and an equal-weight friend occupy the seats; it remains balanced if you and your friend are each joined by another person, as long as their weights match and they sit equidistant from the fulcrum.

In fact, Archimedes continues, the center-of-gravity proposition holds for any number of additional weight-pairs placed equidistantly along the lever. Thus, *the overall center of gravity of an entire system of weights spaced uniformly along a lever lies at the midpoint of the line connecting the centers of gravity of the innermost pair of weights*. Let's call this Archimedes' corollary. Figure 5-3 illustrates this situation for six equal weights arranged uniformly along a lever. That the overall center of gravity remains at the fulcrum can be seen by considering successive pairs of weights labeled 3 and 4, 2 and 5, 1 and 6. Each of these pairs has its center of gravity at the fulcrum; their combined center of gravity falls there as well.

With his logical foundation in place, Archimedes now proceeds to prove the law of the lever, $D/d = w/W$. The proof, which applies to any combination of weights, is most easily illustrated for a specific case. Let's assume that the magnitudes of the weights are in the ratio of 1:3, or $w/W = \frac{1}{3}$, as in figure 5-4a. Archimedes' goal is to prove that balance is achieved when the corresponding ratio of the distances D/d is likewise 1:3.

Considering the 1:3 ratio of the weights w and W, Archimedes divides the lever into $1 + 3 = 4$ equal segments, labeled s_1 through s_4

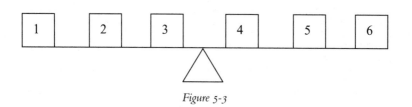

Figure 5-3

in figure 5-4b. Next he slices the heavier weight W into $2 \times 3 = 6$ equal pieces, each weighing $W/6$. He places three of these pieces into segments s_1, s_2, and s_3, to the right of W's former location (figure 5-4b), and the three remaining pieces into corresponding segments s_5, s_6, and s_7, newly added to the left of W's former location (figure 5-4c). True, with these latter segments, Archimedes has augmented the lever. But his lever is abstract and weightless; adding to its length, in itself, doesn't affect the balance properties of the lever. For the moment, segment s_4 remains empty. (Had the ratio of weights been, say, 1:5, Archimedes would have divided the lever instead into $1 + 5 = 6$ segments, sliced W into $2 \times 5 = 10$ pieces, placed five of these pieces to the right of W's former location, and the remaining five into added segments to the left.)

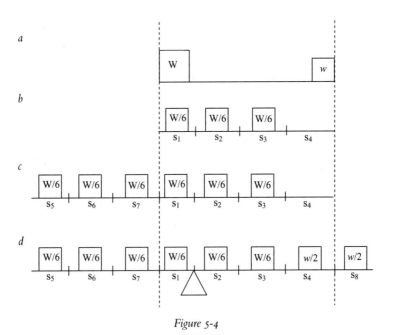

Figure 5-4

Next Archimedes slices weight w into $2 \times 1 = 2$ equal pieces, each weighing $w/2$, and redistributes these in analogous fashion: one piece into segment s_4, to the left of w's former location, and the other piece into segment s_8, added to the right (figure 5-4d). We now have altogether eight pieces of the original weights strung uniformly along the lever. By Archimedes' earlier corollary, the six pieces carved from W, each weighing $W/6$, together have the same center of gravity that W originally did; taken as a system, the six pieces exactly mimic W's lever-tilting capability. Likewise, the two pieces of w, each weighing $w/2$, together have the same center of gravity that w did; they effectively replace w in the problem. Whatever conclusion Archimedes draws from his transformed lever applies to the original one.

Now the rabbit is about to pop out of the Archimedean hat. If weights w and W are in the ratio of $1{:}3$, then W is equivalent to $3w$, which means that $W/6 = w/2$. In other words, each of the six pieces carved from W weighs as much as each of the two pieces carved from w; all eight pieces sitting on Archimedes' transformed lever are identical in weight. The original asymmetrical two-weight lever has become—fanfare!—a symmetrical eight-weight lever.

By symmetry, the balance point of the system must lie at the midpoint of the eight equal weights, where segments s_1 and s_2 are joined. This point is one segment-length away from W's former location and three segment-lengths away from w's; the distances of the original weights are indeed in inverse proportion to the specified magnitudes of these weights. The law of the lever is proved. Of course, Archimedes proves the law for any ratio of weights, not just for the case described here.

Archimedes studied only the static behavior of the lever. He had nothing to say about the dynamics of levers, that is, he is silent on the nature of the forces that cause them to tilt or to balance. That was to follow almost two millennia later in the wake of Isaac Newton's groundbreaking work on force and motion. Still, Archimedes' proof is a masterful application of deductive geometry applied to a physical device.

If there was truly a Herculean test of Archimedes' knowledge of levers and pulleys, it was the launch of what may have been antiquity's largest ship, the *Syracusia*. Commissioned by King Hieron as a gift for Egypt's ruler, Ptolemy, the *Syracusia* weighed in at more than two thousand tons and was said to have had a library, gymnasium, bath, chapel, promenades lined with flower beds, a multicolored floor mosaic depicting the entire story of the *Iliad*, plus accommodations for hundreds of passengers, soldiers, and horses. Archimedes himself saw to it that the vessel was equipped with advanced weaponry of his own design, as well as an Archimedean screw to bail the bilge.

Once built, the ungainly ship needed a shove into the sea. To the king's amazement, so the story goes, Archimedes contrived a mechanical system of compound pulleys, levers, or screws, with which he single-handedly launched the vessel while sitting at ease some distance off. Speculation swirls around the details of Archimedes' device and whether he, in fact, moved the *Syracusia* itself or another ship or merely a heavy object which, over time, morphed into the *Syracusia*.

Like his exploits in ship launching, Archimedes' invention of the irrigation screw that bears his name is disputed by historians. Some contend that he merely learned of it while studying in Egypt as a young man. Yet no such device appears to have existed before Archimedes' time there, and it would have been a reasonable offspring of his keen interests in geometry and engineering. Motor-driven forms of the Archimedes screw are employed in modern pumping stations and wastewater treatment plants. A tiny version is also found in mechanical cardiac assist systems, which maintain blood flow in patients with heart failure or undergoing heart surgery. The screw conveyor, a variation of the Archimedes screw, is commonly used in industry to move grains, powders, and other bulk materials through troughs or pipes.

Of all his mechanical wonders, Archimedes may have had particular fondness for his planetarium; it is the only one of his inventions about which he wrote. Unfortunately, the treatise that describes its making is known only by name through the works of others. Sources from antiquity, including Cicero, who saw the apparatus, describe an extraordinary mechanical engine with spheres representing the Moon, Sun, and five known planets, all circulating at their proper relative speeds around the central Earth, as the universe was conceived back then. In some fashion, the ersatz Moon passed through its monthly phases, and both it and the Sun experienced eclipses. The entire mechanism is said to have been enclosed within a star-studded glass sphere.

In the eyes of the ancients, the evidence was plain. Archimedes had created his own working universe, coaxed water uphill through a screw, moved a ship effortlessly across dry land. His intellect was surely closer to that of a god than to that of a man. And any who doubted his seemingly divine powers had only to witness the horrors unleashed when those faculties were later applied to the art of war.

Chapter 6

THE SCIENCE OF FEAR

*Such terror had seized upon the Romans, that, if they
did but see a little rope or a piece of wood from the wall,
instantly crying out, that there it was again, Archimedes
was about to let fly some engine at them, they turned
their backs and fled.*

—PLUTARCH, *MARCELLUS*

THE ABSTRACTED SCHOLAR who scaled the heights of
geometry, conjured challenging entertainments for kings and
entertaining challenges for colleagues, and felt the shock of lightning inspiration in the bathtub, also turned his manifold genius to
the blood-soaked arena of battle. Indeed, toss Archimedes' name at
an ancient or medieval intellectual, and the knowing response, a
tally of the sage's accomplishments, would begin with his weapons
of war. For many centuries, the popular Archimedes was a military
figure first, a clever inventor second, and a brilliant mathematician
not at all—at least not until his works became widely available
during the Renaissance.

The mythology surrounding Archimedes is equally strong in

the military realm as in the scholarly or practical. To boast of moving the Earth is a charming fiction; to overturn battleships with giant mechanical arms, hurl boulders at advancing vessels, unleash tempests of arrows from dark slits in a wall, pluck men from the field with outsize fishhooks—is to channel the power of the gods. And yet, all true. Archimedes was simultaneously defense secretary, five-star general, and one-man Skunk Works, devising advanced weaponry and formulating, if not implementing, strategy.

As crude as ancient battle might appear to modern eyes, with its arrows and boulders and swords, Archimedes' role in the Roman attack on Syracuse previews the later development of scientific warfare. In any age, Archimedes shows us, it is possible for one side to gain advantage over the other not by virtue of numbers but by harnessing new technology—and, in Archimedes' case, by exerting the most effective weapon of all: fear.

During Archimedes' lifetime, Syracuse straddled a perilous geopolitical fault line between Rome and Carthage. In 263 B.C., a year into the First Punic War, the Syracusan ruler Hieron II assessed the relative strengths of the two superpowers and shifted the city's alliance from Carthage to Rome. The cost for switching sides was an annual tribute of money and grain to Rome plus the enmity of Carthage, neither of which affected the city's ongoing prosperity. When the First Punic War ended in 241 B.C. with a Roman victory, Syracuse reaped a decades-long period of stability and peace.

Hieron was much beloved among the Syracusan populace during his fifty-four-year reign. The ancient historian Polybius observes that Hieron "acquired the sovereignty of Syracuse and her allies by his own merit, having found ready provided for him by fortune neither wealth, fame, nor anything else. And, what is more, he made himself king of Syracuse unaided, without killing, exiling, or injuring a single citizen, which indeed is the most remarkable thing of all; and not only did he acquire his sovereignty so, but maintained it in the same manner." When Hieron tried to resign at an advanced age, citizens urged him to remain in office. In compromise, he established his son, Gelon II, as co-regent.

Hostilities between Rome and Carthage resumed in 218 B.C., when Hannibal marched over the Alps into Italy, the opening salvo of the Second Punic War. Hannibal's early successes emboldened Carthaginian sympathizers within Syracuse to voice their opinion that the city end its decades-long subservience to Rome. Hieron was no stranger to war or to politics. He knew that the enmity between Rome and Carthage placed Syracuse in grave danger from one or the other of these superpowers. He commissioned Archimedes to update the city's defenses. Dionysius's wall still stood, as did the Euryalus fortress, twin bulwarks against outside attack. But siege techniques had improved markedly since Dionysius's time, almost two centuries before. Tales abounded of walled cities falling to persistent enemies. For the besieged, it was only a matter of time before a gate was splintered or a wall breached or the population starved into submission. And then the horrors truly began.

When he was a young officer, Hieron had certainly heard tales of the Macedonian general Demetrius Poliorcetes, Stormer of Cities, who attacked Rhodes with his giant siege tower, Helepolis. "The base of it was exactly square," Plutarch writes, "each side containing twenty-four cubits; it rose to a height of thirty-three cubits, growing narrower from the base to the top. Within were several apartments or chambers, which were to be filled with armed men, and in every story the front towards the enemy had windows for discharging missiles of all sorts, the whole being filled with soldiers for every description of fighting. And what was most wonderful was that, notwithstanding its size, when it was moved it never tottered or inclined to one side, but went forward on its base in perfect equilibrium, with a loud noise and great impetus, astounding the minds, and yet at the same time charming the eyes of all the beholders." And that was now century-old technology, Hieron surely surmised. Who knew what engines Rome or Carthage might throw at Syracuse in the coming years?

As to the steady improvement of catapults, a Syracusan invention adopted by both the Romans and the Carthaginians, Hieron had only to look at his own Euryalus fortress. His predecessor, Agathocles, had added to Euryalus five towers atop which was laid a continuous wooden platform. Each tower held a large, long-range catapult, with four smaller, intermediate-range catapults in the intervening spaces. The firing platform stood some forty feet above the ground, giving the Euryalus catapults a significant range advantage over the enemy's ground-level artillery. In terms of sheer firepower, the Euryalus fortress was a land-based battleship and secured Syracuse's western flank. Nevertheless, Hieron might have grumbled, this battleship was fixed in stone. If the enemy attacked by land from the north or south or by sea from the east, Euryalus would be of no help in the defense of the city.

Archimedes took his commission from King Hieron seriously. In the years leading up to the eventual Roman attack on Syracuse in 213 B.C., Archimedes developed a comprehensive defense strategy, including responses to attacks by land or by sea, by day or by night, from a distance or directly upon the walls. The efficiency and lethality of Syracusan weapons were enhanced according to Archimedes' store of practical knowledge. Ammunition was stockpiled where most needed. To keep away enemy artillery batteries, ditches were dug and barriers erected in front of the city gates and the Euryalus catapult tower. (The placement of the outermost ditch indicates that the Euryalus catapults could hurl a boulder almost six hundred feet.) And every defender knew precisely the task for which he was responsible and had been trained. When the Romans finally came at the city with proven weapons and strategies, the Syracusans stood ready at their preassigned posts, without need to improvise a response.

Hieron and his co-regent Gelon maintained Syracuse's alliance with Rome until 216 B.C., when Hannibal routed the Romans at Cannae. Hieron was now ninety years old, and Gelon may have felt that it was time for his father to step aside. The Roman historian Livy reports that Gelon, "despising his father as a dotard, went over to the Carthaginian interest; and his action would have led to a rising, if, in the very act of arming the populace and trying to get support, he had not been carried off by a death so opportune that even his father did not escape suspicion."

Hieron himself died shortly thereafter, and Gelon's fifteen-year-old son, Hieronymus, a Carthaginian sympathizer whom Polybius describes as "exceedingly capricious and violent," ascended to the throne in 215 B.C. Hieronymus pursued sham negotiations with Rome while simultaneously forging an alliance with Hannibal. He made no secret of his politics, and just thirteen months into his reign, he was assassinated. Hundreds died in the ensuing civil war, in which Syracuse's Carthaginian faction emerged victorious. Two brothers, Hippocrates and Epicydes, both of whom had served in Hannibal's army, became co-rulers of the city. Attack by Rome was now a certainty.

Three major accounts survive of Archimedes' role in the battle for Syracuse. Polybius, who lived from about 200 B.C. to 118 B.C., was closest in time to the action. Greek by birth, Polybius was brought from his home city, Megalopolis, to Rome as a hostage after his family fomented resistance against the Roman occupation. By virtue of his prominent status back home, he was welcomed into Rome's highest social circles and traveled freely throughout the empire. Polybius's sober account of the attack on Syracuse appears in his magnum opus, *The Histories*, which traces the rise of the Roman Empire during the third and second centuries B.C. Of the work's original forty books, only five survive intact, with several others in fragmentary form. From time to time, Polybius interrupts his narrative to editorialize on the writing of history. His coverage of the Syracuse battle is informed by one such observation: "Those who write narratives of particular events, when they have to deal with a subject which is circumscribed and narrow, are compelled for lack of facts to make small things great and to devote much space to matters really not worthy of record. There are some also who fall into a similar error through lack of judgment."

On the heels of Polybius's account is that of Roman historian Titus Livius, or Livy. Born in 59 B.C. into the provincial elite in Patavium, present-day Padua, Livy moved to Rome, where he published his massive *History of Rome from its Foundation* in installments throughout his professional life. Of the original 142 books, just 35 exist today, along with fragments of a few others. Livy's selection of sources and patriotic view of Rome's ascent lead to a sometimes skewed version of events. But his lively writing style and detailed renderings of historical events made him popular in his time.

Of the three major Western historians of the age, Plutarch was farthest removed from the downfall of Syracuse. Born around A.D. 45, Plutarch was wealthy, literate, and schooled in Athens. With his entertaining, but moralistic, *Lives of Noble Greeks and Romans*, he became a favorite on the lecture circuit in Rome. Plutarch had limited interest in the military aspects of the siege of Syracuse, preferring to proffer an ideological screed on the proper societal roles of the scientist (here, Archimedes) and the warrior (the Roman general Marcellus). From Plutarch we gain the impression—probably untrue—that Archimedes disdained practical research and invention, in particular, that he only grudgingly suspended his "purer" mathematical pursuits to engage in the defense of Syracuse.

Compounding and cross-referencing the histories of Polybius, Livy, and Plutarch reveal a fairly complete account of the Roman attack on Syracuse. All three authors agree that Archimedes' military contributions were key to the way events unfolded. Whether Archimedes played an active role during the battle itself is immaterial; his genius for war is evident in the coordinated actions of the Syracusan defenders. Just as he had single-handedly launched a ship, Archimedes leveraged his strategic capabilities through the thousands of hands manning the battlements.

When the Carthaginian faction took over Syracuse—indeed, when all of Sicily appeared to be swinging the "wrong" way—Rome sent one of its foremost generals, Marcus Claudius Marcellus, to quell the revolt and drive the armies of Carthage off the island. Plutarch tells us that Marcellus was battle-tested in both

Gaul and Italy and was "by natural inclinations addicted to war." Perhaps purveying his noble-warrior ideal, Plutarch assures readers that Marcellus was a learned man and venerated Greek thinkers like Archimedes. This same Marcellus ordered the beheading of two thousand residents of Leontini after overrunning that Sicilian city on his way to Syracuse in 213 B.C.

Marcellus knew what he faced in attacking Syracuse. Or at least he thought he knew. Rome and Syracuse had been allies for decades, and the city's fortifications were no secret to the Roman military. In fact, a key element to the effectiveness of Dionysius's wall and the Euryalus fortress was their reputation for impregnability. Marcellus wisely decided to steer clear of Euryalus entirely and concentrate his land forces in a push from the north to Hexapylon, where the wall was less heavily fortified. Simultaneously, his navy would attack from the east and drive defenders from the seawall at Achradina. With this two-pronged assault, Marcellus estimated five days to gain clear advantage over the Syracusans. But as Polybius observes, he "did not reckon with the ability of Archimedes, or foresee that in some cases the genius of one man accomplishes much more than any number of hands."

On the opening day of battle, Marcellus attacked with sixty quinqueremes. These quintuple-oar-bank battleships were plank-for-plank copies of a Carthaginian original captured during the First Punic War, when Rome needed to build a navy in a hurry. Each vessel carried some 270 rowers, 30 crew, plus a contingent of 120 marines—archers, slingers, and javelin throwers—to drive away defenders from the parapets. There were also eight quinqueremes lashed together in pairs, each bearing a sambuca, a shielded ladder which could be thrust atop a wall by pulleys attached to the ships' masts. A screened platform at the leading edge of the sambuca contained soldiers who occupied defenders while their comrades rushed up the ladder to join the fight.

Evidently, Archimedes had studied Roman military methods, for he had readied the seawall with a series of protective devices and strategies. Marcellus ordered his battleships to approach quickly, to minimize their exposure to boulders hurled by catapults. The Romans had expected that, once their ships had approached close enough, the catapults would overshoot them and they would sail unimpeded to the wall. But Archimedes had emplaced banks of catapults with progressively smaller ranges, inflicting damage to the Roman ships during their entire inward passage.

Archimedes is also said to have set ships afire by having his soldiers focus sunlight onto the vessels from banks of mirrors. This is almost certainly untrue. The popular Discovery Channel program *MythBusters* tested a modern version of the supposed weapon, to no effect; however, a group from MIT was able to ignite a stationary wooden target with its own Archimedean "death ray." More reasonably, the Syracusans might have used polished metal plates to reflect the morning sun into the eyes of Roman sailors approaching from the east. Burning mirrors or not, the devastation wrought by Archimedes' catapults alone convinced Marcellus to turn his ships and flee.

To avoid the deadly hail of catapult stones, Marcellus ordered a night attack. Under cover of darkness, the fleet did arrive safely at the Achradina wall. But once there, the ships' decks were raked by bolts and razor-sharp darts shot through rows of loopholes in the wall by unseen "scorpions"—ancient crossbows. Archimedes had tapered the loopholes' interior walls—wider on the inside, a mere palm's width on the outside—to give archers maximal target coverage with minimal exposure.

As the deadly hail of projectiles emerged from the darkness, wooden cranes swung over the parapets and dropped boulders, beams, and lead weights onto the ships below. Plutarch vividly portrays the destruction of one Roman sambuca: "There was discharged a piece of rock of ten talents weight [about six hundred pounds], then a second and a third, which, striking upon it with immense force and a noise like thunder, broke all its foundation to pieces, shook out all its fastenings, and completely dislodged it from the bridge."

Most fearsome of all was the device that came to be known as Archimedes' claw, the mutant offspring of Archimedes' mathematical studies of levers and the instability of floating paraboloids. Essentially a modified cargo-loading crane, the device had a counterweighted boom that carried on its business end a chain-held grapnel—an iron hand, as Polybius describes it. Operators behind the wall swiveled the beam out over the sea, dropped the grapnel, ensnared a ship, then swiftly lifted the ship's prow clear of the water—"a dreadful thing to behold," Plutarch tells his readers. By some means, never described by Archimedes, the claw shook the ship violently, terrifying even the most stolid among the crew, before dropping the vessel back into the sea or dashing it against the coastal rocks. Built narrow for speed, the Roman quinqueremes inevitably swamped or capsized.

In the darkness and confusion, the Romans must have been horrified at the din of battle: whizzing arrows, groaning cranes, crashing ships, shrieking men. They had never before come up against weapons and methods such as these, wielded with such potency by out-of-sight hands behind walls. Could these truly be the machines of men? No wonder, Plutarch relates, "the Romans, seeing that indefinite mischief overwhelmed them from no visible means, began to think they were fighting with the gods." Once again, Marcellus turned back, acknowledging to his troops that the architect of his defeat, Archimedes, had rendered assault by sea impossible.

Meanwhile, the land attack was going no better. Archimedes had ensured that here, too, Roman siege engines and soldiers would be decimated by the incessant bombardment of catapults. Plutarch describes the scene in its full horror: "When Archimedes began to ply his engines, he at once shot against the land forces all sorts of missile weapons, and immense masses of stone that came down with incredible noise and violence; against which no man could stand; for they knocked down those upon whom they fell in heaps, breaking all their ranks and files."

Soldiers who did make it to the wall were unable to set up their scaling ladders in the withering fire from hidden scorpions. Boulders and beams rained down on the Romans, and the dreaded Archimedes' claws hoisted men by their armor or their flesh before flinging them to the ground. Marcellus's land commander, Appius Claudius Pulcher, convened a council of his tribunes, who voted unanimously to end their waking nightmare and cease attack.

Faced with this "geometrical Briareus," as Plutarch dubs Archimedes, recalling the hundred-armed giant of myth, Marcellus realized that any direct assault on Syracuse was futile. Archimedes had foreseen every avenue of attack and, for each one, had prepared a defense. Archimedes proved himself more than just a clever engineer who used levers and pulleys to military advantage. He was a skilled defensive tactician, as evidenced by the remarkable efficiency with which the various defensive elements were brought to bear in the heat of battle. The defenders had been trained to act as a unified fighting entity. "The Syracusans," Plutarch aptly notes, "were but the body of Archimedes' designs, one soul moving and governing all; for, laying aside all other arms, with this alone they infested the Romans and protected themselves." Marcellus's options had been reduced by Archimedes to one: lay siege to the city and starve the residents into submission.

Historians differ as to the hardships endured by the Syracusans. Marcellus sealed off the main roads to Syracuse and stationed warships in the harbor. Nevertheless, the blockade of the city was porous. Galleys were needed all over Sicily, and Carthaginian vessels sailed with impunity through the stretched Roman line. Carthage also dispatched a sizable land force—twenty-five thousand infantry and three thousand cavalry—to harry Roman troops ringing Syracuse.

If the frustrated Marcellus beseeched the gods for a lucky break, they delivered in the form of one Damippus, a captured Spartan envoy sent by Syracuse to engage the support of King Philip of Macedon. The Syracusans sought Damippus's release and offered to negotiate on neutral ground outside the northern wall. During the talks, a Roman officer noticed what appeared to be an unmanned tower, a promising entry point for a small force. He surreptitiously estimated the tower's height by counting the courses of stone from the ground upward. Entry lay within reach of a scaling ladder.

The gods appear to have favored Marcellus a second time, for a Syracusan deserter wandered into the Roman camp and alerted them to the upcoming festival of Artemis. If past experience was any guide, the city's celebration would begin early and continue through the night, with free flow of wine to boost residents' spirits. Everyone would partake—including the guards.

On festival night, Marcellus marched his troops silently toward the gate at Hexapylon. About a quarter mile to the east, a small raiding party ascended ladders into the tower the Romans had previously scouted. The soldiers stole along the parapets, overpowered the few remaining guards at Hexapylon, and opened the gates for their compatriots. Marcellus's army poured onto the Epipolae Plateau, only to find it subdivided into fortified townships. The path to Syracuse through Achradina was blocked by another wall, this one actively defended. And to Marcellus's rear, ready to discharge its garrison against him, stood the Euryalus fortress.

Word of Marcellus's subterfuge reached Carthage, which dispatched an armada of one hundred ships plus a sizable land force that encamped in the marshy lowlands southwest of Syracuse. Once again, fate favored Marcellus. Plague swept through the Carthaginian army, virtually wiping them out. Unfavorable winds and Roman vessels kept the Carthaginian navy from reaching the harbor. After two years of siege, Syracusan spirits were at an ebb. Marcellus attacked in force. This time, the city surrendered.

The Roman soldiers, who had long suffered the torments of battle against the Syracusans, wanted to burn the city to the ground. Marcellus refused. After posting guards on the treasury and on the homes of Roman sympathizers, he permitted his men to sack the city, but to harm no one—especially not Archimedes. Marcellus would deliver into Roman hands not only the city of Syracuse but the living embodiment of Greek scholarship himself. Never mind that this one man had inflicted so much pain upon his soldiers. And never mind that he, Marcellus, routinely lopped off the heads of vanquished generals and kings. Archimedes, whom the gods had endowed with insight verging on their own, was a prize best kept alive.

With Syracuse fallen, Archimedes returns to center stage to enact the final scene of his epic life. Ancient historians and their medieval successors all penned versions of what happened to Archimedes in 212 B.C. as the Roman army swept through the city. From the perspective of these writers, for whom accuracy might be compromised to make a point, Archimedes was simply too important a figure to exit in less than noble fashion. And so, another Archimedean tale arose that would become an intellectual talisman to medieval and Renaissance thinkers. True or not, here it is: Archimedes is oblivious to the ransacking of the city. He is immersed in thought, hunched over some geometric figure he had traced out in a tray of sand. A Roman soldier bursts into his dwelling, perhaps unaware of the identity of the old man before him. The soldier advances upon Archimedes, intent only on securing gold or jewels, surely indifferent to valueless traces in the sand. But Archimedes, drawn out of his mathematical reverie, is defiant and scolds the intruder, "Do not disturb my circles." Or perhaps, as another writer reports, Archimedes says, "Stand away, fellow, from my diagram." Or, "Fellow, stand away from my line!" Or maybe Archimedes says nothing at all and merely waves the man away. Whichever the scenario, if any of these, the enraged soldier brings down his sword and ends the life of antiquity's greatest mathematician.

Marcellus was stricken with the news of Archimedes' murder. According to Livy, he arranged a grand funeral in the necropolis outside the city's Achradina gate. A monument was erected on the burial site, in keeping with Archimedes' expressed wish: a stone marker engraved with the figure of a sphere circumscribed by a cylinder; and below, an epitaph announcing the perfect 2:3 ratio of the surface areas of the nested shapes. Archimedes—to some, an inventor; to others, an engineer; to the Romans, a mythic opponent; to himself, ever the geometer.

Part II

A PALIMPSEST'S TALE

Chapter 7

THE VOICE BENEATH THE PAGE

When I get a little money I buy books; and if any is
left I buy food and clothes.

—Erasmus

L IKE AN ITINERANT traveler, the timeworn manuscript
  crisscrossed the Byzantine Empire, borne over ancient byways
by caretakers oblivious to the secret harbored within its mildewed
leaves. To the monks entrusted with its care, the manuscript was a
Euchologion—an Orthodox Christian liturgical manual—written
in Greek by an anonymous hand in the dim recesses of history.
Page through it, and you would find, among other things, blessings
for the loaves at Easter; a genuflection for Pentecost; the naming
ceremony for a newborn child; rites for engagement, marriage, re-
pentance, sickness, and death; readings for the principal feasts; even
exhortations to cleanse a contaminated well or a vessel of rancid
olive oil.

In this leg of its long, peripatetic existence, the manuscript
journeyed from a monastic library in the Holy Land to the
Metochion—"daughter-house"—of Jerusalem's Church of the

Holy Sepulchre, in Constantinople. And there, during the summer of 1906, it fell open under the gaze of Professor Johan Ludvig Heiberg, the world's foremost philologist, who had traveled
in haste from Copenhagen to read the aged document. With his
biblical beard and riveting gaze, Heiberg's very presence cast
the library into a confluence of the ancient and modern worlds.
He was, by vocation, a conduit who funneled the long-ago into
the present-day by parsing words in spotted, crumbling, centuries-
old documents. Heiberg's particular passion was ancient Greek
geometry. Over the past three decades, he had scrutinized every
line of every extant work of every known Greek mathematician
from antiquity. He had published dozens of books and papers de-
tailing these ancient masterpieces of logic. His mind was suffused
with geometric propositions and proofs, each one as stylistically
characteristic of its particular author as fingerprints. By this point,
so far into his career, Heiberg could be forgiven if he believed he
had seen everything there was to see of ancient Greek mathematics. And yet, unless his practiced eyes were somehow askew, the
document spread wide before him might be, as he was hoping, the
discovery of a lifetime.

The manuscript took the form of a smallish book—a codex—
177 leaves in all, as unremarkable looking as a pebble by the side
of the road. Its Greek script, inked in somber brown strokes,
burst forth now and then in pale flourishes of red and blue, as
though the scribe's artistic anima had wrested control of the pen.
The voice of the long-departed transcriber intrudes only briefly,
on page 75, where he entreats the Lord to remember him. But no
name is penned there.

A skilled paleographer, Heiberg quickly surmised from the
characteristic style of the lettering that the text dated from

the late twelfth century. It was indeed a liturgical document, as he had been told. Yet neither its great age nor its content held any particular interest for him. Driven by winds of rumor, hope, and experience, Heiberg had crossed a continent in pursuit of a document far more important than the Euchologion. The work he sought conveyed ideas whose genesis lay not in the twelfth century but more than a thousand years earlier. A work that had been lost to history. A work whose scholarly value, if recovered, was inestimable.

Heiberg rotated the codex a quarter turn. The words of the Euchologion, now oriented vertically down the page, became a mere veil of meaningless symbols, beyond which Heiberg peered into a murkier, more distant stratum of time. Here, barely visible even to Heiberg's trained eye, were lines of Greek, running at right angles to the overlying liturgical text: the ghostly remnants of a previous document. Heiberg peered through his magnifier. The slightly cursive *theta*, the enlarged *upsilon*, the truncated stroke of the *psi*—the scribe who had penned the Euchologion had no part in this. The underlying text was clearly in a different hand.

Heiberg knew precisely what had happened. Around the twelfth or thirteenth century, when the Euchologion was likely authored, paper—a Chinese innovation—was not yet widely available in Byzantium; important documents were written on vellum, a fine parchment made from the skin of a calf. Lacking a supply of the costly vellum, the Euchologion scribe chose to erase the codex's older work—literally scrub it out of existence—to create a tabula rasa for his new commission. Had he taken the trouble to scrape the pages with a blade, rather than merely wash them with a sponge, the vellum might have returned to its pristine state.

Fortunately for Johan Heiberg, the scribe's efforts were insufficient to his task; the existing letters, rendered in durable oak-gall ink, clung to the page like so many limpets to a rock. Afterward, the scribe cut down the vellum sheets, turned them sideways so remnants of the old writing now ran unobtrusively down the page, inked the Euchologion over its predecessor text, then rebound the sheets. The book became a *palimpsest*—from the Greek *palimpsestos* for "scraped again"—a recycled manuscript containing multiple texts superimposed.

Heiberg ran his magnifying glass over the palimpsest's phantom script. The faded strokes formed letters, the letters formed words: σημεῖον, point; γραμμή, line; κύκλος, circle; σφαῖρα, sphere; κύλινδρος, cylinder. Heiberg's solemn eyes must have widened with excitement as he confirmed that this was no ordinary text but a collection of exquisitely sophisticated scientific and mathematical treatises, accompanied by fine-lined geometrical diagrams throughout the narrative. And what a narrative it was. Heiberg had seen many ancient Greek technical works that unfolded in the rigid axiomatic style of Euclid. And while the palimpsest had plenty of that, its fraying pages also contained lengthy sections in which the author speaks directly to the reader, providing a window onto the logical ruminations of an extraordinary mind.

Although some of the passages were unfamiliar to Heiberg, there was no mistaking the characteristic Doric dialect, much less the incredible sophistication of the mathematics, unique in the ancient world. And once Heiberg's magnifying lens revealed the titles of the treatises—*On the Equilibrium of Planes*, *On Floating Bodies*, *Measurement of a Circle*—any lingering doubts about authorship disappeared. These could only be the works of Archimedes, the greatest scientific sage of antiquity.

Heiberg paged through the book, sifting his fingers through what he realized was scholarly treasure—indeed, the find of a lifetime. There were seven treatises in all, including one, the *Method of Mechanical Theorems*, whose existence had been inferred only through references in other ancient works. Archimedes' famous analysis of buoyancy appeared here, too—in its entirety, and in the original Greek, not the idiosyncratic Latin translation long known to the world. There was even a brief passage describing Archimedes' take on an ancient puzzle called the Stomachion. Altogether, the palimpsest formed the earliest extant collection of Archimedes' treatises and was arguably the most important ancient scientific document discovered in the modern era. In this modest-looking book, Heiberg found writings pregnant with interpretive possibility, writings whose very structure and creative expression, like a symphony or poem, might reveal the mind of its author.

At this point, Heiberg might have put down his magnifier, leaned back, and considered his good fortune—or more appropriately his professional obligation as intermediary between the ancient and modern worlds. The Turkish authorities had resisted his efforts to inspect the manuscript, at one point even denying its very existence. Yet here it was, gloriously real. The palimpsest was all he could have hoped for—and more. Chance had burdened him with a weighty responsibility. He had to rescue the Archimedes Palimpsest before it again faded from human memory.

"Where the Turk's foot is planted, grass never grows again," the *New York Times* reminded its readers on July 16, 1907. Yet, in seeming defiance of that dour adage stood the page-one announcement that, during the previous summer, Copenhagen's Johan Ludvig

Heiberg had uncovered a series of long-forgotten works of Archimedes in a monastic library in Constantinople. Heiberg's find plus the revelation that the sultan's library contained some three thousand ancient manuscripts ranged in leather cases upon the wall led the *Times* to a hopeful conclusion: Perhaps Islamic invaders had *not* burned all of Constantinople's treasured antiquities when they overran the depleted metropolis in 1543. In particular, libraries may have been spared, perchance some of their manuscripts contained the name of God. Lost works of Livy or Cicero might have escaped the ancient conqueror's bonfires. Or one of Heiberg's counterparts, inspecting a dusty basement shelf, might stumble upon the poems of Sappho, which the breathless *Times* report pronounced "the greatest literary loss the world ever suffered."

Farther down the column, Meredith Townsend, former editor of London's weekly newspaper, the *Spectator*, and author of *Asia and Europe*, demanded a thorough search of the crypt of St. Sophia. The search for literary treasures, Townsend huffed, should begin "before the great day when the destiny of the Ottomans is completed, and Constantinople once more sinks down, a mass of blood-stained ruins, fired by its possessors before they commence their final retreat to the desert from which, in the mysterious providence of God, they were suffered to emerge, in order to destroy the eastern half of the civilized world."

In the summer of 1906, a year before Townsend's heated rhetoric, Johan Heiberg was quietly worrying, not about the sultan's books or hypothetical lost jottings of Livy, but about his career-topping work with the Archimedes Palimpsest. His sole concerns at the moment were the preservation and transcription of one of the most important mathematical manuscripts ever written. And time was running out.

The same government functionaries who had earlier denied the palimpsest's existence had just turned down Heiberg's second request to borrow it for further study. Short of settling in Constantinople, there was no way he would have enough time to copy the manuscript, much less transcribe it, before summer's end when he had to return home to Copenhagen. He had tried to explain to Turkish officials the daunting task that lay ahead of him. This was, after all, no ordinary document. The twelfth-century scribe who had ripped apart the original mathematical codex had shuffled the reconstituted pages before overwriting the prayer book, forcing Heiberg to collate the unnumbered Archimedean pages solely on their contents. Even for Heiberg, a three-decades-plus veteran of ancient Greek mathematics, this would be a time-consuming job. To complicate matters, the ancient sage's words had been partially sponged away, if not completely obscured by the overlying text. And essential parts of the old document lay hidden within the stitching of the book's spine. The cover would have to be removed and the binding undone to bring this unseen text to light.

Only then, having made out the writing on the page, could the real work begin. Word division, an innovation of the ninth century, had not yet reached the scribe who drafted the Archimedean text; Heiberg would have to parse strings of letters into words, words into sentences, sentences into paragraphs. The letters of the day had multiple forms and ligatures, confounding interpretation. Abbreviations and word contractions were everywhere, many of them nontraditional; evidently, the Archimedean scribe was trying to conserve vellum. In short, Heiberg's mission was to eliminate the host of textual peculiarities and outright errors that would render a verbatim copy of the palimpsest all but incomprehensible to the modern-day reader.

With the clock ticking—and with the possibility that the fickle Turkish bureaucracy might restrict future access to the manuscript—Heiberg petitioned for permission to photograph the palimpsest. This time the response was positive. Conveniently for Heiberg, Constantinople had become a world center for photography. Tourists eagerly paid for images of exotic Ottoman street scenes or for portraits of themselves dressed as a Turkish pasha. Photographers from all over Europe had descended on Constantinople, typically staying a few years, then moving on.

To decipher the palimpsest, Heiberg required photographic images with exceptional contrast and definition. Even inspecting the original pages with a magnifying glass, he could barely make out the faded text. Nine of the manuscript's leaves were completely illegible, and several others revealed of Archimedes no more than a few scattered words. Heiberg knew that the photographs would have to render intact every detail of the centuries-old lettering and the hairline geometrical diagrams. Anything less would do him no good once he was back in Copenhagen.

Without delay, Heiberg hurried down the now-familiar Grande Rue de Pera in Constantinople's posh European quarter and entered the storefront at number 414. Here was the studio of seventy-year-old Swedish-German émigré Guillaume Berggren, who had established his reputation with masterful portraits of Anatolian life, architecture, and historical sites. At the time, Berggren was riding high off European commissions, tourist photographs, and now Heiberg's unexpected assignment to photograph a timeworn book. Yet a mere decade later, as World War I bled the Turkish economy, Berggren would survive by selling his precious glass negatives for use as windowpanes.

The palimpsest photographs completed, Heiberg returned to

Copenhagen, where he painstakingly identified, collated, and transcribed the underlying Archimedean works. (Sixty-five of the photographs currently reside in an album in Copenhagen's Royal Library; Heiberg's identification of the various treatises is written in shorthand on the back of each image.) He revisited Constantinople during the summer of 1908 for a final inspection of the original manuscript.

It must have been with a pang of regret that Johan Heiberg closed the Archimedes Palimpsest in the library of the Metochion before heading back to Copenhagen in 1908. For two years, he had scrutinized its faded pages, both in photographs and now, for what would prove to be the last time, in reality. Every visible pen stroke, crease, and blemish of the palimpsest had come under his magnifying glass, until he knew its parchment terrain as intimately as that of his home city. The fragile codex marked the very pinnacle of Heiberg's career as a classicist. In a way, his decades of single-minded devotion to the study of antiquity had been prelude to his culminating work on the palimpsest. Arguably, there was no one else in the world at the time who could have done it as well.

Walking the streets of Constantinople that day in 1908, Heiberg might have reflected on the serendipity of his situation: that the winding thread of the palimpsest's peripatetic existence should have crossed his own. More than two thousand years earlier, Archimedes' treatises were recorded either by him or by a scribe on papyrus scrolls. Copies of these scrolls were routed to colleagues in Alexandria, who commented on them and may have disseminated additional copies. These early-generation scrolls had long since atomized with age. For his latter-day rendezvous with the palimpsest, Heiberg was indebted to a succession of unheralded scribes who had labored over centuries to copy and re-copy worn Archimedean treatises. The palimpsest represented the effort of one such scribe, whose pen strokes had endured nearly seven hundred years before coming under Heiberg's eye. How far into the future might the palimpsest's life line extend? Centuries, millennia? Or months, weeks, perhaps only days. Heiberg knew well the sad inventory of lost works of antiquity. *Lost.* The catchall euphemism for forgotten, erased, decayed, burned, cut up, stolen.

Even basking in the afterglow of his time with the resurrected palimpsest, Heiberg could only lament the loss of Archimedes' other works—on polyhedra, optics, calendars, construction of planetariums—not to mention the agonizing gaps in the rest of humanity's written heritage. As Heiberg had learned, were it not for the reference by Archimedes in his work *The Sand-Reckoner,* no one would know that Aristarchus had anticipated the Renaissance-era Copernican model of the cosmos by some eighteen centuries. Of Archimedes' illustrious predecessor, Pythagoras, not a single text survives. Of the 123 plays of Sophocles, a mere 7 remain, and of Euripides, just 18 out of more than 90. With few exceptions, Sappho's poems are gone. Strabo's *History* has likewise vanished.

Heiberg had done everything he could to record for posterity the writings of Archimedes. But in every generation are too few like himself who strive to preserve humanity's birthright through its ancient texts. Now he had no choice but to abandon his cherished codex to the whims of time, nature, and human folly.

In its long existence in the Middle East, the palimpsest had been a silent bystander to incessant religious, tribal, and nationalist conflicts. Only a decade after Johan Heiberg returned to Copenhagen in 1908, the codex was hunkered down in Constantinople, the Eastern epicenter of a sprawling European conflict. The Great

War was a showpiece of deadly technology: tanks, planes, poison gas, submarines. A military engineer himself, Archimedes might have marveled at the continued inventiveness of humanity to subdue and destroy its own. In his day, he had developed machines of war every bit as frightening to his ancient foes as were modern-era armaments to twentieth-century soldiers. Paradoxically, this same architect of destruction elevated the human spirit by creating incomparable works of science and mathematics, several of which now lay half-erased, half-forgotten, and vulnerable in a defeated city.

With the dissolution of the Ottoman Empire after World War I, Turkey found itself on the receiving end of a punitive peace settlement, which placed the country under Allied occupation and left its economy slave to other nations. The internal struggle between the ruling sultanate and the nationalist government only added to the turmoil. Johan Heiberg never returned to Constantinople to reconnect with the palimpsest. If he had, he would have been heartsick. Reaching toward the shelf in the Metochion where once had stood a smallish book of particular interest, Heiberg would have found an empty space. By the 1920s, the Archimedes Palimpsest was gone.

Chapter 8

A BRIDGE ACROSS TIME

The palest ink is better than the best memory.

—CHINESE PROVERB

T<small>O THE UNTRAINED</small> eye, nothing is more impenetrable looking than the wrinkled pages of an ancient manuscript. Yet to classicists like Johan Heiberg, these pocked, sometimes moldy leaves are invaluable treasure, a literary armature around which the flesh of history can be applied. The archaeological detritus of a vanished civilization sketches the face of ancient life and accomplishment; a more vivid portrait materializes with the recovery of its written works. Often writing is the most potent extant witness to the varied societal glue that once bound together a people. Where writing is absent, the historical record is inevitably dimmed.

That latter-day mathematicians have been able to retrace the logical paths trod by Archimedes hinged on the invention of writing. Had Archimedes not recorded his ideas in a written language nor had the means to do so, the fruits of his genius would have evaporated with his passing in 212 B.C. Incredibly, the contents of his long-vanished letters to Alexandrian colleagues survive, having run a veritable gauntlet of destruction, wrought by the frailty of the writing medium itself as well as by human thickheadedness. To the degree that remnants of Archimedes' original hand linger in these mathematical missives, something of the man, not just the myth, can be glimpsed among the tangle of ancient characters.

The Western world's rediscovery of Archimedes during the Middle Ages is rooted in the history of writing and, by extension, in the preservation and transmission of ancient manuscripts and derived texts. Civilization and writing appear to go hand in hand. The ancients scribbled characters on all manner of surfaces: wooden slabs, bark, slate, clay or wax tablets, walls, even pottery shards— *ostraca*—the Post-it notes of antiquity. With the growing complexity of human interactions arose the need to dodge the vagaries of shared memory, whether to record literature, laws, natural history, commercial and legal transactions, family trees, or the works of Archimedes. Vainglorious rulers, from the dawn of civilization onward, deified themselves on monuments, à la Shelley's Ozymandias. Plebeians likewise used ancient walls to eternalize their humble existence or to covertly scrawl what they dared not voice in public. In every age, in every place, prophets and disciples painted their various visions through the written word.

Sumerian scribes, from southern Mesopotamia, were among the Western world's earliest professional writers, active in both commercial and literary contexts. They wrote by impressing wedge-shaped characters—cuneiform—into soft clay tablets with a stylus. These language-bearing tablets were hardened in the sun or in kilns, making them extremely durable. A Mesopotamian text from, say, 1700 B.C. is practically as readable as the day it was created. Yet an effective writing system requires more than the proper materials and personnel; it calls for a shared symbolism that is widely understood within a population and not too taxing on the writers themselves. Not only did the medium of clay limit the speed with which scribes could write; it mired the Mesopotamians in an unwieldy writing method with hundreds of complex, linear-angular characters. Tedium must have been the watchword for these ancient scribes, who had to make multiple impressions with the stylus to render each character. For all its durability, clay discouraged the development of a more rapid curvilinear or cursive script, the portal to wider dissemination of written documents.

That Mesopotamian scribes were key to the kingdom's overall vitality is affirmed by the 150,000 cuneiform clay tablets excavated from the region, dating as far back as the fourth millennium B.C. Although these ancient copyists were viewed by the populace as specialized clerks, they apparently felt a degree of ownership of their writing; one Mesopotamian colophon contains both a blessing upon readers who preserve the text and a curse upon those who deface or erase it. The education given these scribes was valued by at least one long-ago parent, who wrote his son's teacher, "My little fellow has opened his hand, and you made wisdom enter there: you showed him the fine points of the scribal arts." To be a scribe in ancient Mesopotamia was a laudable, if not lucrative, profession.

From Egypt has come down to us the first record of writing for sale: An enterprising undertaker offered *The Book of the Dead* at funerals, dividing the proceeds between himself and the priests. The scribes who created such texts are ubiquitous in ancient Egyptian paintings and sculpture. Some stand ramrod straight, displaying the implements of their trade: slate palette, ink cups, reed brush. Others sit cross-legged, brush poised, traditional scribal kilt stretched tablelike between the knees. Each stares at the viewer with canine alertness, a sentient writing appliance awaiting input. Royal figures, too, are depicted with palette and brush, reflecting the high regard ancient Egyptians must have had for the written word and, presumably, for those who wrote. Tutankhamen himself departed for the afterlife with writing implements tucked into his tomb.

Egyptian scribes climbed higher on the social ladder than their Mesopotamian counterparts. Notaries, copyists, and especially personal secretaries in Egypt could accumulate great wealth and influence. One royal scribe, Horemhab, rose to pharaoh following the death of Tutankhamen's successor. (It didn't hurt that Horemhab's résumé included a stint as general of the Egyptian army.)

Egyptian scribes trained individually or in court schools, where promising youngsters learned the essentials of administration within the highly centralized government. Four thousand years ago, the Egyptian bureaucrat Dua-Khety advised his school-bound son, Pepy, to be diligent in his scribal studies: "It is to writings that you must set your mind . . . I do not see an office comparable with [the scribe's.] . . . I shall make you love books more than [you love] your mother."

Once at school, Pepy might have longed for his mother's love. The pedagogic model was rooted in memorization, repetitive practice, and corporal punishment. Students copied form letters, literary texts, and other instructional documents. They committed to memory the forms and nuances of some seven hundred pictographic symbols, later named *hieroglyphs*—sacred writing—by Greek invaders. For the recalcitrant student, there was also regular application of the hippopotamus whip. "The ear of a boy is on his back," a surviving account declares, "he hears when he is beaten." Notably, the ancient Egyptian word for teach—*seba*—also means "beat."

Had there been an ancient equivalent of the 1960s film *The Graduate*, the word whispered into the young protagonist's ear would have been *papyrus*. As a writing medium, papyrus was the "plastics" of its age: lucrative, abundant, and high-tech—well, at least compared to Mesopotamian clay. Egypt held a virtual monopoly on sheets made from the papyrus reed—the plant didn't grow anywhere else—and exported it throughout the Mediterranean from the third millennium B.C. onward. Papyrus was costly and fragile. Yet for lengthy documents, there was no practical alternative at the time; all that scribal ink had to flow somewhere. When ancient mathematicians read Archimedes, they were most likely reading from papyrus.

The papyrus plant was once ubiquitous in marshes along the Nile. (With the advent of wood-pulp paper, papyrus was no longer cultivated; a modest stand was reintroduced near Cairo in 1969.) To preserve its Mediterranean-wide monopoly, Egypt kept the papyrus-making process secret. Our best guess—confirmed by modern experiment—is that the plant's stalks were cut into narrow strips, which were overlaid along their edges. A second layer was arranged crosswise, and the combination was pressed, dried, and rubbed smooth until it cohered into a single sheet. Since papyrus is brittle and cannot be creased or folded, individual sheets were glued with starch paste into long scrolls.

The ends of each scroll were affixed to sticks or bars, which were turned together to advance through the document. Scribes typically wrote in a series of columns, several of which could be viewed at a time. Only the inner side of the scroll was used; writing on the backside would have been scuffed away with repeated unfurling and furling of the scroll. There was typically no title page in an ancient scroll; the narrative literally unwound from the very beginning. The colophon, if present, containing essential information about the text—author, title, date, occasionally the scribe's name—typically appeared at the end of the scroll. Apparently, having read the text, the reader merely left the scroll for the next reader, who opened it up to the identifying colophon.

Pity the poor scholar who wished to locate a passage buried somewhere within a scroll. There was no index, nor even the possibility of an index, to assist in finding the selection. The reader had no alternative but to unfurl the scroll and scan the columns. As the Alexandrian poet and librarian Callimachus grumbled in the third century B.C., "A big roll is a big nuisance." To allay the problem, Callimachus literally cut up epic works, such as the 150-foot-long *Odyssey* or the *Histories* of Herodotus, into separate scrolls, or "books."

Scribes wrote on papyrus with a reed brush-pen, about eight inches long, one end either crushed or slant-cut, depending on the particular application. Black ink was created by mixing water with a dark powder, such as soot, crushed charcoal, or lampblack, a by-product of heating carbon-rich oils. Red ink, which was commonly used for titles and initial letters, was pigmented with cinnabar, a crystal ore of mercury, or minium, a lead compound. Gum arabic, from the viscous sap of the acacia tree, was often added to prevent the ink from running.

While Egypt successfully exported its writing medium, its cumbersome writing system did not find such a ready market. Egyptian hieroglyphic was no less complex than Mesopotamian cuneiform.

On the paperlike papyrus surface, hieroglyphic was as creaky as a Model T on a superhighway. Therefore, scribes reserved the sacred symbolism for formal inscriptions in tombs and temples and developed a simpler, hieroglyph-inspired cursive script, called *hieratic*, for writing on papyrus. Around 700 B.C., an even simpler and faster writing system arose, which the Greek historian Herodotus referred to as *demotic*, or "writing of the people." At first, demotic was adopted for civil records, but soon spread to religious and literary documents. The Rosetta Stone, which famously opened the door to the decipherment of ancient Egyptian writing, contains hieroglyphic, demotic, and classical Greek versions of the same text.

A key breakthrough in the evolution of writing arose in the Middle East before 1000 B.C.: the alphabet. Whereas a given hieroglyph conveys a particular idea—one symbol for, say, baboon, and another for the Sun—each alphabetic letter represents a pronunciation element of the spoken language. Thus letters are untethered by meaning and can be strung together to form a diverse array of words. The advent of alphabetic text reduced the number of symbols needed for written expression from many hundreds of pictographic elements to a couple of dozen letters. This not only made writing and reading easier, but opened the door to more widespread literacy.

Archimedes and his Greek contemporaries would have used an alphabetic writing system inherited from commerce-savvy Phoenician traders in the eighth century B.C. (An earlier Mycenaean writing system fell out of use on the Greek peninsula around 1100 B.C. upon the downfall of the Mycenaean civilization.) The original Phoenician script did not serve the Greek language completely. Like Hebrew and other Semitic languages, spoken Phoeni-

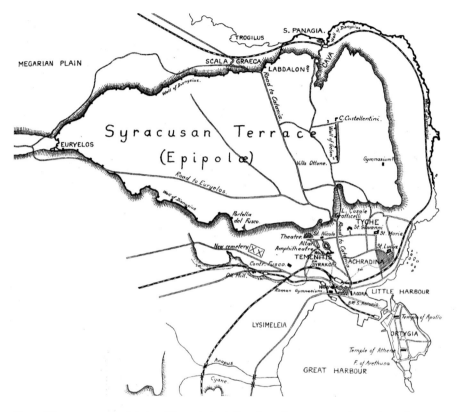

Map of Syracuse and vicinity. The original Greek settlement was situated on the harbor island of Ortygia, at lower right. Portions of the defensive wall of Dionysius still enclose the Epipolae plateau, with remnants of the Euryalus fortress to the west. (*Fabricius [1932], by permission, Deutsche Archäologische Institut*)

Ruins of the Euryalus fortress, which once guarded the western approach to Syracuse. Remnants of the catapult tower are seen at top. (*Photograph by John Dean*)

Possible design of the Archimedes' Claw, used by Syracusans during the Roman siege of 213 B.C. (*From* Ancient Inventions *by James & Thorpe, Ballantine Books, 1994. Illustration by Peter Koenig*)

Johan Ludvig Heiberg, photographed around 1918. (*Julie Laurberg, Royal Library, Denmark*)

Mar Saba monastery, east of Bethlehem, as depicted in the book *Picturesque Palestine*, 1881. (*Picturesque Palestine/www.LifeintheHolyLand.com*)

Cicero Discovering the Tomb of Archimedes, painted in 1804 by Benjamin West. (*Yale University Art Gallery*)

The Archimedes Palimpsest, as it appeared in 1999, (above) closed and (below) open to a proposition from Archimedes' lost treatise, the *Method of Mechanical Theorems*. Only the overlying religious text is visible. (*Photograph by John Dean*)

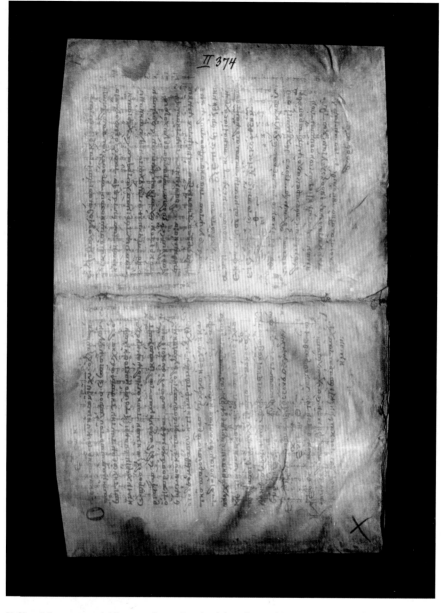

Folios 16 verso and 17 recto from the Archimedes Palimpsest, photographed in 1906. Here, Euchologion text runs vertically, perpendicular to the Palimpsest's spine; Archimedes' *On Floating Bodies* runs horizontally, parallel to the spine, in two columns. (*The Digital Palimpsest, http://www.archimedespalimpsest.org/*)

Again, folios 16 verso and 17 recto from the Archimedes Palimpsest, now digitally imaged and enhanced to reveal text from Archimedes' treatise *On Floating Bodies*. Several geometric diagrams straddle the book's spine in the right-hand column. (*The Digital Palimpsest, http://www.archimedespalimpsest.org/*)

cian was largely consonantal. Its alphabet had no vowels, which made accurate vocalization of written Greek difficult. As a result, the Greeks recast a number of Phoenician consonants as vowels: *A* (alpha), *E* (epsilon), *I* (iota), *O* (omicron), and *Y* (upsilon). Vowels *H* (eta) and *Ω* (omega) came later. Well before Archimedes' time in the third century B.C., the Greek alphabet comprised a mere twenty-four letters—seventeen consonants and seven vowels—each with uppercase and lowercase forms. The Greek alphabet spawned, among others, the Etruscan alphabet in central Italy, which itself gave rise during the seventh century B.C. to the Latin alphabet, the source of many modern European languages.

With its slimmed-down script and the Hellenic imperative to widen the realm of ideas, Greek writing took off. By the time Archimedes came on the scene, papyrus scrolls abounded throughout the Mediterranean, with a steady stream of intellectual output impressed upon the page. Yet, while the written record was expanding, its long-term preservation was a haphazard affair. Papyrus decays. After a time, many original works became too fragmented to decipher. The papyrus bridge that might have served as our written link to the ancient world was falling down even as it was being erected.

Chapter 9

THE PARCHMENT BROTHERS

*If you do not know what writing is, you may think it
is not especially difficult . . . Let me tell you that it is
an arduous task: it destroys your eyesight, bends your
spine, squeezes your stomach and your sides, pinches
your lower back, and makes your whole body ache . . .
Like the sailor arriving at the port, so the writer rejoices
at the last line.*

—Colophon of a twelfth-century
religious manuscript from
northern Spain

THE ANCIENT HELLENIC empire, with its host of commercial and literary enterprises, offered ample opportunities
for creators of new ideas as well as for the professional scribes
whose task it was to write them down. In Athens, a philosophy lecture by Aristotle's noted student Theophrastus drew an audience
of two thousand. Monarchs competed to attract scholars to their
courts. Royal patronage supported scientists, poets, playwrights,
and mathematicians.

The buzz of scholarly pursuits generated a burgeoning manuscript trade. Big-name philosophers spared no coin to buy precious scrolls for their private libraries. Plato handed over three Attic talents—the present-day equivalent of seventy-five thousand dollars—for a trio of works by Philolaus. Aristotle paid just as dearly for writings of Plato's nephew, Speusippus. Scribal services were likewise needed at institutional libraries that dotted the Mediterranean, most notably at Alexandria. Prices for custom writing, from legal documents to love letters, were set by decree: so many lines for so much money, with a bonus for superior penmanship. Individual scribes developed a reputation for the beauty of their script or the accuracy of their transcription.

All of this feverish writing activity hinged on one thing: sufficient papyrus to take up the flood of ink. During the second century B.C., if the Roman historian Varro can be trusted, a monopolistic Egypt refused to supply papyrus to its rival Pergamon in Asia Minor. In desperation, the scribes of Pergamon are said to have improved the venerable animal-skin writing surface. Whether the story is true or not, parchment—from the Greek *pergamene*, "skin from Pergamon"—became the first true competitor of papyrus. By the time Archimedes' treatises were set down in the palimpsest during the tenth century, papyrus had largely left the scene in favor of parchment (as parchment would some five centuries onward when paper arrived in the West).

To make parchment, the hides of sheep, lambs, or calves are softened in a lime bath, then stretched on a rectangular frame. Flesh and hair are scrubbed away with a knife blade and pumice stone. The skin is dusted with chalk or similar powder to absorb leftover oils, then scraped to remove any remaining blemishes and irregularities. The result: a uniform, nearly white sheet that takes

ink well. A superior form of parchment called *vellum*, made from
the skin of very young or even stillborn calves, was created for
special volumes, such as those with brightly colored illuminations.

Parchment production in the West took off with the introduction, around the second century A.D., of the codex, the forerunner
of the modern book. As a codex medium, parchment is superior to
papyrus in every way. Not only can it be inked with a goose quill,
which is more durable than a reed pen and requires less dipping
into the ink well, but its strength and texture permit writing on
both sides. Parchment leaves can be produced to larger dimensions
than a papyrus sheet, and they remain intact when folded for binding, whereas papyrus is apt to split.

The codex itself, whether of papyrus or parchment, has significant advantages over the scroll. It is compact yet capacious enough
to hold multiple texts. (The Archimedes Palimpsest codex contains
seven geometric treatises plus several other unrelated documents.)
The codex is far easier to consult for specific text, a property of
immense importance to early Christians, who needed quick access
to scriptural passages. And, if constructed of parchment, a codex
withstands the ravages of time better than a papyrus scroll.

The rise of the codex during the early Christian era forced
the selective conversion of classical documents from scrolls into
codices. With each codex requiring a veritable flock of animal
skins—the 392-leaf Souvigny Bible consumed around two
hundred—there was never enough parchment to accommodate
more than a tiny percentage of ancient works. Most of the works
that didn't make the cut from scroll to codex inevitably decayed
into dust. And even ones that were converted might later be
washed away and re-inked with a document adjudged more im-

portant. Now and then, the washing was incomplete and a ghost of the old text poked through—a palimpsest.

After the Roman conquest of Greece, the center of book production shifted away from Athens, first to Alexandria, and then to Rome itself. High-volume publishing in the Roman Empire was accomplished by the scribal equivalent of galley slaves: pools of learned drudges who simultaneously penned copies of the exemplar work from dictation. During the second century A.D., the number of written works produced in the Roman Empire doubled compared to the century before. Business was brisk for both Greek classics and original Roman literature.

Rome was astir with literary activity. Poets recited at banquets. Authors read their latest works at public forums. Travelers of means carried a book to read aloud to themselves on the journey. While overall literacy remained limited within the empire, there was Hellenic-inspired elementary education for children to learn language, reading fundamentals, and arithmetic. Privileged students continued their education with private tutors, who guided them in oratory and the study of classical Greek works.

Such were the Roman reader's tastes and so time-consuming and labor-intensive was the copying of scrolls—in essence, publication—that, only a few centuries after their creation, specialized works like the treatises of Archimedes became difficult to obtain. This despite the fact that there were at least twenty-eight libraries within Rome itself and numerous provincial libraries scattered across the empire. With the rise of Christianity and with Constantine's decision around A.D. 324 to move the capital of the Roman Empire to Byzantium, the nexus of book production began to shift away from Rome. The third century saw one-third fewer literary and scientific books published than the second, and the fourth century two-thirds fewer than the third. By the time Rome was overrun by the Goths in 410, the preservation of classical literature and science already rested in the hands of the empire's Christian monasteries.

The monastic scribe's original order of business was to produce copies of various liturgical guides for religious practice, such as the Euchologion in the Archimedes Palimpsest. Some of these documents remained in-house; others made their way to affiliated abbeys. During the sixth century, Italian monasteries founded by the Roman statesman and writer Cassiodorus at Vivarium and by St. Benedict at Montecassino spearheaded a more literate approach to religious devotion and the scribal arts. Cassiodorus set down detailed guidelines for a *schola Christiana* where, in addition to the requisite study of Holy Scripture, monks would read the likes of Hippocrates on medicine, Strabo on geography, and—explicitly named—Archimedes on geometry. Cassiodorus extolled the virtue of making worldly knowledge available through the copying and dissemination of manuscripts. To him can be credited the gradual transformation of Christian monasteries into the de facto publishing houses and literary repositories of the medieval Western world.

If Cassiodorus envisioned the monastery as a religious institution of letters, his contemporary, St. Benedict saw it as essentially the opposite: a place where secular studies augmented one's service to God. To Benedict, learning for learning's sake was an affront. Benedict's Rule, so-called, declared that monks read for three hours each day during the summer and two hours during the winter. They were further compelled to complete an entire book during Lent and to pack a book whenever traveling. This at a time when reading was considered the mental equivalent of labor in the fields; given the freedom, many a monk would have chosen the hoe over Herodotus. A team of senior monks stood by to ensure that Benedict's reading directive was carried out. At the conclusion of Lent, each monk was questioned about the contents of his assigned book. If his answers were deficient, he was sentenced by the abbot to another term with the book. The librarian saw to the distribution of the monastery's treasured texts and, most critically, to their return. According to medieval commentaries, a missing book triggered a monastery-wide search by the abbot himself.

A second element of Benedict's Rule promoted daily oral readings to the brethren during meals and assemblies. Given the limited literacy of monks at the time, this was probably more effective than individual reading in teaching classical works. So critical were these periods of group edification that the gathered monks communicated with one another by hand signals, else a voice interrupt the reader. The reader's task was arduous: enunciating syllable by syllable, as was the mode of the time, a cramped Greek or Latin text, devoid of word spacing or punctuation. Benedict advised the reader to take some bread and wine beforehand to keep from flagging during the long recital. Apparently, readers were held in high regard within the monastic community; the title was inscribed as an honorific on their tombstones.

The Benedictine practice of reading aloud to oneself or to the group spread to other monasteries. (Silent reading was uncommon until the twelfth century, although some eight centuries earlier St. Augustine reported of his contemporary, St. Ambrose, that "his eyes ran over the page and his heart perceived the sense, but his voice and tongue were silent.") In 790, Charlemagne declared Benedict's Rule of study to be an obligatory part of religious devotion and practice, stimulating the reconsideration and rediscovery of classical works, including those in science and mathematics. Scholarly scribes of the Carolingian era took an activist approach to translation, correcting obvious transcription errors and resolving inconsistencies in the text. The most authoritative editions of a work would receive the official imprimatur *ex authentico libro*.

The Carolingian era also saw a dramatic expansion of monastic libraries. By the end of the ninth century, the Benedictine monastery at Bobbio, in northern Italy, housed almost 700 religious and secular codices, said to be the largest collection in Europe at the time. Even this paled in comparison to Byzantine state and church libraries: The imperial library of Constantinople, as early as 475, contained 120,000 volumes, including a prized 120-foot-long parchment scroll of the *Iliad* and the *Odyssey*. So prevalent did monastic libraries become by the twelfth century that one cleric observed, "A cloister without bookcases is like a military camp without armament." This monastic tradition of scholarly study would lead, in part, to the survival of the works of Archimedes.

During the Middle Ages, reproduction of manuscripts became a major source of income for monasteries, as wealthy patrons sought copies of rare works for their libraries. The beleaguered cartoon character Dilbert might have found the environment depressingly familiar: monks scribbling away in their respective carrels—a medieval variant of the office cubicle—egged on by the pointy-hooded (pointy-haired?) boss, who was typically also the monastery's choirmaster. Scribes sat or stood at their desks, laboring throughout the day to copy up to four large leaves—eight pages—of text. Sunset brought a reprieve: To ensure the safety of the precious manuscripts, firelight of any kind was banned. If business was good or if a commission arrived for multiple copies of a work, the monastery might take up Cassiodorus's suggestion that an entire room—a scriptorium—be allocated to the task. The scriptorium would have been one of the few heated spaces in a monastery complex and, of necessity, would have been suffused with natural light. In some cases, a reader probably dictated the exemplar work to one or more copyists (which might account for the frequent spelling errors found in medieval manuscripts). Otherwise, scribes mumbled their respective texts to themselves, infusing the medieval scriptorium with a sound one observer likened to the hum of a beehive.

These anonymous monastic copyists were told by the abbot which texts to copy and often had little or no understanding of what they were writing. Their attention was focused—if focused at all—on the letter, not the word. Monks were to be humble movers of the pen, copying their assigned works verbatim. For their efforts—and the inevitable eyestrain, headaches, and hand cramps—they received neither public acclaim nor stipend. The creation of a book was service to God and to the order, an act of piety, a manifest expression of one's religious devotion. The scribe's sole reward: self-satisfaction.

Yet these monks were merely men and, piety be damned, were apt to complain—in writing, of course. "St. Patrick of Armagh," an Irish scribe pleads in a marginal note, "deliver me from writing." Another declares, "Thank God, it will soon be dark." The colophon of one manuscript dispenses with the usual bibliographic details and contains instead, "Now I've written the whole thing: for Christ's sake give me a drink." And for the potential book abuser, this overtly un-Christian take on the old Mesopotamian curse: "For him that stealeth, or borroweth and returneth not, this book from its owner, let it change into a serpent in his hand and rend him. Let

him be struck with palsy, and all his members blasted. Let him lan-
guish in pain crying aloud for mercy, and let there be no surcease to
his agony till he sing in dissolution. Let bookworms gnaw his en-
trails in token of the Worm that dieth not, and when at last he
goeth to his final punishment, let the flames of Hell consume him
forever." Pop these sentiments into the mouths of today's librarians,
and books would be returned on time.

Inevitably, there was the occasional scribal monk to whom hu-
mility was foreign. Take the twelfth-century's Eadwine of Canter-
bury, self-declared "prince of writers," whose portrait annotation
rings with a Maileresque bravado undiminished by time: "Neither
my fame nor my praise will die quickly . . . Fame proclaims you in
your writing for ever Eadwine." And rarely, a human spark erupts
playfully from the page, like this marginal jot by a ninth-century
Irish monk: "Pleasant to me is the glittering of the sun today upon
these margins, because it flickers so." Yet whatever their personal
faults or devotional shortcomings, medieval scribes were esteemed
within the Church; the penalty in seventh-century Ireland for
killing a scribe was the same as for murdering a bishop.

Euclid wrote for the geometric education of the masses, virtually
ensuring the survival of his works through the ages. Archimedes,
on the other hand, wrote for the relatively small number of mathematical specialists like himself. The transmission of Archimedes'
works therefore depended on continued demand by a meager
cadre of highly trained scholars, both in the West and in the Islamic world. During times of political, cultural, or religious up-
heaval, when advanced learning was discouraged, the number of
Archimedean patrons dwindled to near-extinction. Already barely
a generation after Archimedes' death in 212 B.C., the Greek mathe-

matician Diocles was unable to procure a copy of his famous predecessor's *On the Sphere and Cylinder*. Many ancient commentators sang the praises of Archimedes' mathematical works, without ever having laid eyes upon them. In fact, we know more today about the breadth of Archimedes' mathematical treatises than did most mathematicians of antiquity.

The writings of the ancient Greeks survive today because of the sporadic emergence of societal oases where, for a few precious decades, the scholarly tradition of the classical era flourished. It was in one such oasis—medieval Byzantium—that the works of bygone scholars like Archimedes were transcribed from decaying exemplars and bound into books. But the Byzantine enlightenment gradually dimmed, and the once-prized books found fewer and fewer readers. Like a fleet of parchment time machines, these vessels of classical knowledge were cast onto the shifting seas of memory and circumstance. Only a relative few arrived in safe harbor, into the light of study; the rest foundered into obscurity, taking with them the fruits of a glorious age.

Three compilations of Archimedes' works were generated in Byzantium. All three left their Byzantine birthplace, effectively orphaned after their patrons died. The books passed through various hands, many driven by the pride of possession, almost all dumb to the mathematical insights offered them. To the innumerate owners of these volumes, the name Archimedes was largely an emblem that endowed their fragile ream of parchment with value. Of the three Byzantine compilations, two are lost. The third survives but, in the utmost of ironies, only through a singular act. Having outlived its usefulness as a mathematics text, the book was erased.

Chapter 10

LEO'S LIBRARY

Apart from a handful of works preserved by papyri on the arid desert fringe of Egypt, every Greek text that we read today owes its survival to Byzantine copyists.

—ROBERT BROWNING,
"BYZANTINE SCHOLARSHIP," 1964

THE MISSIVE FROM al-Ma'mun, caliph of Baghdad, must have come as a shock to Leo, if the old Byzantine anecdote is true (which it is probably not). Why would the exalted ruler of the Muslims and enemy of the Byzantine people wish to communicate with *him*, a hovel-dwelling teacher of mathematics in ninth-century Constantinople? What confluence of astrological signs had Leo neglected to discern that ordained such an unlikely intersection between the lofty and the low? In pursuing the root of al-Ma'mun's interest, Leo would have come to understand that he and the caliph were joined less by cosmic action than by earthbound forces: war and geometry. The message was, after all, an invitation for Leo to appear in the caliph's court. And if Leo found that honor alone insufficient to secure the deal, there was the offer of substantial gold—an amount Leo, a mathematician, undoubtedly computed—that exceeded a lifetime's income from teaching.

If there was one thing al-Ma'mun prized beyond power, riches, or territory, it was the dream of turning Baghdad into the Alexandria of its day: the nexus of the world's intellectual activity. He had thrown his considerable power behind Baghdad's Bayt al-Hikmah—House of Wisdom—a library and academy to which were brought the literary and mathematical works of the world's scholars—and often the scholars themselves. In al-Ma'mun's eyes, the Byzantine people were doubly deficient: Not only were they infidels, they were the withered offspring of once-glorious forebears. Who were the more rightful heirs to the ancient wisdom—the great Islamic thinkers or the culturally benighted Byzantines? The caliph's goal was to reconstitute in Arabic all of classical Greek learning and, from there, explore new intellectual realms. (Like its Alexandrian predecessor, the House of Wisdom declined under a later orthodox Islamic regime, and was destroyed entirely in the Mongol invasion of Baghdad in 1258. The Tigris is said to have run black with the ink of books hurled into its waters.)

As the old tale continues, the caliph tapped Leo to lead his kingdom's charge into the frontiers of mathematics (this despite the fact that the House of Wisdom already employed vaunted mathematicians like Muhammad al-Khwarizmi, one of the founders of algebra).

Leo might have wondered how his name had come to be known in faraway Baghdad. The caliph's courier would have recounted how a particular Byzantine soldier, seized in battle, had dazzled his Muslim captors with his mathematical prowess, only to reveal that he was but the pupil of a much greater mathematician—his teacher—in Constantinople. This mathematical master, the soldier continued, had crisscrossed the Byzantine Empire in pursuit of the great works of the ancients: Euclid, Aristotle, Archimedes. He had painstakingly copied each treatise and, although poor, possessed a personal library the caliph himself might envy.

If Leo had swooned at al-Ma'mun's offer, he soon recovered. In his hands lay a private communiqué from the leader of the Muslims, a man responsible for the deaths of many of Leo's countrymen. Lest he be accused of consorting with the enemy, Leo is said to have rushed the caliph's message to a government minister, who brought it directly to Theophilus, emperor of the Byzantines. Theophilus was a man of culture, who was keenly aware of Baghdad's obsession with classical Greek science, mathematics, and philosophy. With Greek manuscripts ubiquitous in the Arab East, al-Ma'mun threatened the preeminence of Constantinople as steward of the ancient intellectual tradition. Now, with his proposal to Leo, the caliph was tapping the very rootstock of Byzantine scholarship.

As Theophilus well knew, his battles against the caliphate had been disastrous of late. Yet here was an instance where he might secure victory without dispatching a single soldier to the eastern front. Against al-Ma'mun's gold, he reportedly offered Leo a state salary and an appointment as a teacher in the Patriarchal School at the Church of the Forty Martyrs.

Leo accepted Theophilus's offer. In this state-sanctioned post, he would have both the authority and the means to expand his collection of classical treatises. He could gather and copy not just samples of his predecessors' works but complete collections. The treatises of Archimedes, say, which by the ninth century were so difficult to find, might be assembled in one or two volumes, pre-

serving this precious legacy for generations to come. From alley-
way to the emperor's side, the purported geometry of Leo's career
was one Archimedes would have appreciated: an ascending arc.

In Archimedes' time, and for centuries thereafter, the most complete collection of Greek scholarly manuscripts was housed at the
Library in Alexandria, where Archimedes himself had once studied
and where copies of his treatises must have lain among the many
scrolls. Domination by Rome, followed by rejection of so-called
pagan Greek works by Rome's Christian successors, reduced the
intellectual hub of Alexandria to a shadow of its former self. The
once great Library, already depleted of its works, went up in smoke
during the Islamic invasions of the seventh century. Divorced from
its cultural patrimony, the West became an intellectual backwater.
Both the fruits of Hellenic scholarship and the Hellenic learning
tradition evaporated from the consciousness of the continent's inhabitants, as they struggled to survive the economic and political
hardships of the day. Only in the monasteries did any semblance of
scholarly study continue. Starting with the disintegration of the
Roman Empire in the fifth century and lasting the better part of a
millennium, much of western Europe wallowed in the Dark Ages.

   At the same time, scholars in the Islamic world, which by the
seventh century stretched throughout the Mediterranean region,
were translating ancient Greek treatises into Arabic. They added
their own observations and reflections, although religious traditions
prevented them from significantly advancing the state of science
from what it was in Archimedes' era. These Arabic translations
would eventually find their way to European centers of learning,
where, once translated into Latin, they would nourish long-starved
minds.

The Islamic nations were not the only ones with an interest in antiquity. Already by the sixth century, with Alexandria in decline, the center of Greek scholarship was shifting to Byzantium, the Christian eastern remnant of the old Roman Empire, founded two centuries earlier by Constantine. The Eastern Church saw itself as steward of the Hellenic classical tradition, responsible for bearing that torch into the future. Classical manuscripts that would have otherwise suffered extinction in Alexandria were copied and shipped off to the Byzantine capital, Constantinople, there to be studied and recopied. As a result, virtually our entire original-language legacy of classical Greek arts and science—the Archimedes Palimpsest included—stems from medieval Byzantium.

Greek was the official language of the Byzantine Empire, and nearly everyone in the government, from the lowliest bureaucrats to the monarchs themselves, learned to emulate the expressive style of their Hellenic predecessors. Education was integral to Byzantine society, the better to provide a continuous stream of literate bureaucrats. Aspiring civil servants might have looked to section 14 of the Theodosian Code of laws: "No person shall obtain a post of the first rank unless it shall be proved that he excels in long practice of liberal studies, and that he is so polished in literary matters that words flow from his pen faultlessly."

The highly centralized government administered a host of taxes, budgetary matters, and complex regulations. Laws, official documents, and state records were written in literary Greek, which by this time was substantially different in vocabulary, grammar, and pronunciation from the Greek used in everyday discourse. Essentially, Byzantine writers had to learn a foreign variation of their own language. To top it off, the Byzantines clung fast to the same

Greek alphabetic number system that Archimedes' contemporaries had struggled with centuries earlier.

It's no accident that the word *Byzantine* has come to mean "excessively complicated." The scholars of the Eastern Empire gained a reputation for obscurity, emphasizing form over content. The dense prose of their bureaucratic counterparts was equally impenetrable. One exasperated government worker, oblivious to the profound impact the Justinian Code would have on the development of European law, doodled in a tenth-century land assessment manual, "O law, my brain could grasp running water more easily than it could grasp you."

Unlike medieval Europe, where the scholastic pipeline was controlled by ecclesiastics, elementary education and literacy were widespread among the Byzantine populace, even in rural areas. Students started off with basic instruction in the Greek alphabet and language, then took up reading to learn form and analyze content. This was followed by practice in the art of rhetoric, based on classical prose writers. Last came in-depth study of the philosophy of Plato and Aristotle, with a capstone unit on mathematics, astronomy, and musical theory. With education conducted on an empire-wide scale, adoption of a given classical text into the official curriculum virtually ensured its preservation during that period.

The Byzantine ethos was an uneasy amalgam of Christianity and classical culture. There were always elements within the Byzantine hierarchy who questioned the propriety of pagan teachings. The ongoing debate was effectively summarized by the second-century Christian theologian Tertullian, who asked his fellow scholars, "What has Athens to do with Jerusalem? What has the academy to do with the church?"

Byzantine intellectuals found much in classical Greek philosophy and ethics that was concordant with Christian theology. Works of pagan authors were freely published and used widely for classroom instruction in grammar and style, although sometimes accompanied by withering denunciations of their content. In this culturally schizophrenic environment, the impiety of the classical Greeks was roundly condemned while their artistic merits were simultaneously extolled. Basil the Great, a popular Christian intellectual, wrote a treatise in the fourth century titled *Exhortation to Young Men as to How They Shall Best Profit by the Writings of Pagan Authors*. Scripture became the framework on which classical Hellenic gloss could be hung. Of course, the mathematics of Archimedes had no contradiction in Scripture.

The Christian historian Socrates wrote in the fifth century: "Pagan culture was neither rejected nor accepted by Christ and his disciples. This was no accident, for many pagan philosophers came near to knowledge of God, and this may be useful to Christians . . . Holy Scripture teaches us admirable and truly divine doctrines, inspires its hearers with piety, uprightness and faith; but it does not teach the craft of letters." In brute irony, ecclesiastical purists freely wielded the standard forms of classical writing to celebrate their own ignorance of classical poetry and philosophy.

The danger in Byzantium was not so much to the pagan works themselves as to the proponents of pagan points of view. It was unwise to admire Plato's philosophy too much in public or outside a purely educational context. More than one scholar was banished from Constantinople for being too "Hellene."

The sixth-century mathematician Eutocius, who studied and later taught at Alexandria, prepared commentaries—essentially, explanatory supplements—to three of Archimedes' treatises: *On the Sphere and Cylinder, Measurement of a Circle*, and *On the Equilibrium of Planes.* Such commentaries were critical to students— teachers, too—who might otherwise be mystified by the omitted steps and leaps of intuition in Archimedean proofs. The fact that Eutocius developed his guide probably means that Archimedes was part of an advanced curriculum at Alexandria. After all, Alexandria is where Archimedes had sent his geometric proofs, to the mathematicians Conon, Dositheus, and Eratosthenes, who saw to their distribution. The commentaries of Eutocius would have been of particular interest to builders of the many edifices commissioned by the renowned Byzantine emperor Justinian during the sixth century. The architects—mathematicians, really—charged in 532 with the reconstruction of the soaring Hagia Sophia church in Constantinople are known to have consulted the works of Archimedes through Eutocius and to have assigned these works to their students.

The reign of the iconoclast emperor Theophilus, from 829 to 842, marked the culmination of a period of economic revival, passable relations with Arab neighbors, unparalleled construction of public buildings, and increased scholarly study and reproduction of classical Greek texts. The empire-wide adoption of the minuscule script, developed by Byzantine scribes decades earlier, made it possible to write more quickly and compactly than before; this smaller, faster writing style allowed an increase in book production, which fostered literacy among the empire's populace. Theophilus himself basked in the reflected glory of Byzantine technocrats, receiving foreign officials in a chamber equipped with mechanized roaring lions and twittering birds.

This same period saw the rise of the Byzantine academic

who came to be known as Leo the Mathematician. It is Leo to whom Renaissance scholars owe their knowledge of much of Archimedes' work. Having found no suitable teacher of rhetoric, philosophy, or mathematics in Constantinople, the young Leo studied for several years with an aged monk on the Aegean island of Andros, then struck out on his own to copy classical works scattered in monastery libraries throughout the Byzantine Empire. Afterward, he returned to the capital to teach privately. In the early 830s came the surprise invitation to visit Baghdad, and in 838, following the Byzantine defeat at Amorium, a second invitation. This time Caliph al-Ma'mun's equally enlightened successor, al-Mu'tasim, dangled before the emperor more favorable terms of peace between their empires. Again Theophilus refused. The intellectual birthright of the Byzantines was not for sale.

Leo was a lifetime collector of books, amassing one of the great personal libraries of the age. Among the library's volumes were writings of Thucydides, Plato, Euclid, and the astronomer Ptolemy, plus works by dozens of other Greek and Byzantine luminaries. If his extant manuscripts are any guide, Leo's own contributions to the mathematical canon are rather feeble compared to those of his illustrious predecessors, and even to those of his Arabic contemporaries. (Perhaps the caliphs of Baghdad were more interested in Leo's library than in the man himself.)

Leo's interests ranged widely—too widely, according to some, who attacked his apparent lack of interest in Christian dogma. To the chagrin of Church officials, Leo openly practiced astrology during his brief stint as archbishop of Thessalonika. He also compiled a well-regarded medical encyclopedia, no doubt from a host of medical books that crowded his shelves. And, echoing Archimedes' own military inventiveness, Leo developed a series of fire beacons between Constantinople and Tarsus, on the eastern frontier, that was used as an early warning system against Arab invasion.

Leo flourished, even as the seat of power and the political winds repeatedly shifted. In his later years, he was named director of a richly endowed school for advanced studies in Constantinople's plush Magnaura Palace. Here he served as mentor to the prolific lexicographer and commentator Photius, who would become known for his *Bibliothecha*, an annotated list of some three hundred books he had read. Through Photius, we gain a sense of the scholarly energy that suffused Constantinople in Leo's time—and of Byzantine efforts to preserve classical knowledge.

With papyrus production at an ebb, demand for parchment in Byzantium far outstripped supply. What could be had was expensive and therefore limited to religious documents, Greek classics, or essential instructional works, such as Euclid's geometry. It would have been easy for Archimedes to fall between the cracks and his remnant works molder into dust. But during the mid-ninth century, presumably under Leo's direction, the dispersed treatises of Archimedes were gathered and copied, then bound into a parchment codex. Dubbed Codex A by Johan Heiberg, this collection was devoted to Archimedes' mathematical works, to which were added Eutocius's commentaries on three of the treatises. A near-contemporaneous compilation, Heiberg's Codex Β, which might also have been connected with Leo, held Archimedean works related to optics and mechanics, including one pioneering treatise, *On Floating Bodies*, that did not appear in Codex A. This may have been the first time so many of Archimedes' treatises had been gathered; at Alexandria, his writings probably existed as individual scrolls or were bound alongside works by others. How crucial were Leo's compilations to latter-day Archimedean scholars? Through meticulous research, Heiberg proved that Codex A and Codex ℬ were the root sources for virtually all subsequent copies and translations into the early twentieth century, when Heiberg published his own translation based in part on a previously unknown collection, his Codex C, the Archimedes Palimpsest.

In a dedicatory note in Codex A from an anonymous scribe, Leo is justly addressed as "dear friend of the Muses." Had Leo not preserved the works of Archimedes during what would prove to be a critical window in Byzantine history, medieval and Renaissance mathematicians would have been limited to a meager fraction of Archimedes' treatises—and these, mediocre Latin translations of previous Arabic editions of primary Greek source documents. Such sequential translation of already abstruse subject matter inevitably leads to misinterpretation and errors; like the game of telephone, what emerges from the end of the line bears only passing resemblance to what goes in.

By the year 1000, Archimedes' legacy rested precariously on three compilations in Constantinople: Leo's two codices plus a third, more recent codex destined to become the Archimedes Palimpsest. Barely two centuries later, in 1204, that legacy found itself in peril when Venetian Crusaders and their allies stopped short of the Holy Land and launched a devastating attack against Constantinople. The armies of the West, united in the Fourth Crusade, stormed through the capital of their Christian brethren, destroying lives, books, and antiquities with utter abandon. The empire's intellectual vigor, especially in technical fields such as astronomy and mathematics, would not recover for another century. True, scholars still wrote, teachers still taught, and the public still read—from this era comes the first reference to the vice of reading in bed. But for a time there would be no more Leos, to whom the mathematics of Archimedes was a soaring example of human accomplishment worthy of preservation. The sublime beauty of Archimedes' geometric theorems became all but opaque to the withered Byzantine intelligentsia. Before the thirteenth century was out, both of Leo's Archimedean compilations would be shipped out of Byzantium. And the third, left behind, would be tossed on the recycling heap—a ready source of parchment and no more.

In the year 1229, a Byzantine scribe set about his task: writing a Euchologion, an essential liturgical guide in the Eastern Orthodox Church. From a monastery library, the scribe had been given a number of surplus codices to yield up their parchment for the new project. One of these codices was unlike the others, filled with mathematical text and geometric diagrams. Whether the scribe noted the name Archimedes on the title page cannot be known. Even if he had, the name would have held no special significance for him. This was, evidently, a text that had outlived its usefulness—a text without advocate. No one was interested in its preservation, least of all an innumerate scribe.

Codex C in its original form was sizable by modern standards: twelve inches tall and almost eight inches wide. Its brown, oak-gall ink text was arrayed in two columns, about thirty-five lines each, with ample margins into which oversize initial letters strayed. The script was of the compact minuscule style, which had become the norm by the tenth century when the text was created. Gossamer-lined diagrams appear throughout, more than fifty in total. In every way—except its contents—this was a typical Byzantine codex of its time.

The destruction of the Archimedean work was a tedious affair.

The Euchologian scribe removed the book's cover, sliced the threads that bound the signatures, and disassembled the parchment leaves. He did the same to other codices before him; the Euchologion would consume more pages than Archimedes alone had to give. Each of the original leaves, with its pair of Archimedean pages straddling the midline, was stretched on a wooden frame. There they were scrubbed with a mild acidic solution, perhaps from citric fruit. Treatise after treatise melted away—*On Floating Bodies, On Spirals, On the Equilibrium of Planes*, and what might have been the sole surviving copy of the revelatory *Method of Mechanical Theorems*—until the old ink became a calligraphic ghost.

After the bleached leaves were dried in the sun, the scribe cut them down the middle. He stacked the halved pieces and trimmed their edges to match—except for one damaged leaf, which he trimmed straightaway without halving it first. When done, he rotated his stockpile of recycled parchment ninety degrees to form the raw leaves for the Euchologion. Each page of Archimedes, oriented sideways, would accommodate two pages of the new work, shrinking the original book down to an almost pocket-size eight inches by six inches. In consequence, each Euchologion page would bear just half an Archimedean page. That the scribe had not even a covert interest in preserving the old work is clear; he used the Archimedean pages out of order and freely interleaved them with pages from other scrubbed works. To reconstruct the original Archimedes manuscript, a reader would have to doggedly pursue a near-invisible mathematical narrative scattered helter-skelter throughout an overlying religious narrative.

The scribe dipped his quill into the inkwell and began to write, crosswise to the old text, the blessing of the loaves at Easter. He proceeded with the liturgy of St. Basil, then psalms to be sung at evening or early morning services, then genuflection for Pentecost. Letter by letter, ink flowed onto parchment, gradually overwriting the treatises of Archimedes. On one leaf, the new writing *parallels* the old—this, the errant sheet the scribe had neglected to cut in half. Here Archimedes' words read literally between the lines, although upside-down.

The Euchologion scribe added an occasional stylistic flourish to the otherwise somber tome. Titles are penned in red. Initial letters are enlarged and bear decorative elements, sometimes in blue. When all done, onto the first page of the Euchologion—and onto a page from Archimedes' *On Floating Bodies*—the scribe wrote his name and the date: *Ioannes Myronas, April 14, 1229.* Completed on Holy Saturday, just one day before Orthodox Easter that year, the Euchologion of Ioannes Myronas would find a ready and grateful audience, none of whom would have the slightest interest in the underlying narrative on its pages. The Archimedes Palimpsest would slumber many centuries before finding a constituency of its own. As pre-Renaissance Europe gradually awakened to the glories of its classical forebears and to its own intellectual vigor, the palimpsest languished incognito on a monastery shelf, awaiting resurrection.

Chapter 11

RESURRECTION AND LIGHT

It has always been my opinion that a mathematician who has not studied the works of Archimedes accurately ought scarcely to be called a mathematician.

—RENAISSANCE MATHEMATICIAN
FEDERIGO COMMANDINO, 1558

AROUND THE TENTH century, some twelve hundred years after Archimedes' death, Europe began its long climb out of the so-called Dark Ages. With growing political stability, improving economic conditions, and the reinvigoration of commerce came a renewed interest in learning. Contact with the Islamic and Byzantine worlds led to the recovery of lost works of antiquity. Ancient texts in Arabic or Greek were rendered into Latin and gradually disseminated throughout Europe. About 1120, the English monk Adelard of Bath disguised himself as a Muslim student and traveled to Cordoba to procure an Arabic copy of Euclid's *Elements*. Adelard's Latin translation became the standard geometry text until 1533, when a fourth-century Greek version of the *Elements* was recovered. During the early thirteenth century

Fibonacci, in Italy, introduced Eastern arithmetic and algebra to Europe, including Arabic numerals and the Hindu-Arabic decimal system.

Archimedes resurfaced in the West in piecemeal fashion. First came a twelfth-century Arabic-to-Latin translation by Gerard of Cremona of Archimedes' *Measurement of a Circle*. The preface to a medieval list of Gerard's translations reveals the dearth of source material available to translators in the Latin West. For his Archimedes manuscript, as well as the seventy-plus other scientific translations he completed during his lifetime, Gerard relocated to the cosmopolitan center of Toledo, which had been retaken from the Muslims. Here he found a multitude of ancient Greek treatises, all rendered in Arabic. Unlike many of his colleagues, who hired intermediaries to first convert the unfamiliar writing into spoken vernacular Spanish, Gerard learned the language to the degree that he could translate directly from the Arabic sources.

The same preface to the list of Gerard's translations conveys the reverence with which ancient scholars were viewed by latter-day Western readers: "As a light shining in darkness must not be set under a bushel, but rather upon a candlestick, so too the splendid deeds of the great must not be held back, buried in timid silence, but must be made known to listeners today, since they open virtue's door to those who follow, and in worthy memorial offers modern eyes the example of the ancients as a model for life."

Sometime during the thirteenth century, in the wake of the Fourth Crusade's assault against their Byzantine brethren, Leo the Mathematician's once-prized Archimedean codices, A and Ⴆ, now each some four hundred years old, were loaded aboard a westbound ship in Constantinople. There was little interest anymore within the Byzantine Empire in ancient mathematics, especially in the esoteric treatises of Archimedes. The codices' destination, appropriately enough, was Archimedes' homeland, Sicily. The recipient: the German royal house of Hohenstaufen, which had reigned over Sicily and southern Italy for decades and transformed the kingdom into a center of culture. It was a time when royal courts throughout Europe vied with monasteries as inheritors of classical scholarship.

Leo's codices lay in the royal repository in Palermo until 1266, when the island's last Germanic ruler, Manfred, was defeated at the battle of Benevento by forces loyal to Pope Clement IV. The codices joined the papal manuscript collection and, in short order, landed on the desk of the Church's foremost Greek translator, the Flemish Dominican cleric William of Moerbeke.

Born around 1215 or later, William of Moerbeke is reported to have studied at Cologne and perhaps at the University of Paris before traveling to the Byzantine city of Nicaea to learn Greek. His earliest translations, during the 1260s, brought the works of Aristotle to St. Thomas Aquinas. From February through December 1269, at the papal court at Viterbo, Italy, William translated Leo's Archimedean codices into Latin. (The original version, in William's own hand, resides in the Vatican library under the title *Ottobonianus latinus 1850*.)

Although not a mathematician, William did identify a number of copying errors in the original Greek manuscripts. He corrected these in his translation and jotted the offending Greek text in the margin, flagged with the word *falsum*. William's marginalia attest to the deteriorated state of both codices. Codex A had faded and lacked a cover. Of the treatise *On Floating Bodies* in Codex B, he complains, "Here in the exemplar there was an empty space of half a folio [one page] and the rest of the demonstration was missing." Many of the diagrams had nearly vanished with age.

Following their translation in 1269, Leo's codices were returned to the papal library. Both are listed among the 645 manuscripts in the library's register from 1311, when Pope Clement V moved his court to Avignon. After that, Codex 𝔅 disappears permanently from the historical record. Its companion, Codex A, left the papal collection as well, apparently bound for private hands.

To the pre-Renaissance reader, Euclid was a mere geometric foothill compared to the rugged pinnacles of Archimedes. Dense with mathematical symbolism, spidery diagrams, and complex reasoning, Archimedes' treatises must have looked virtually impenetrable. In fact, their very existence confounded reason. How could it be that the mathematical architecture of nature had been contemplated so deeply more than a thousand years earlier? Experiencing Archimedes' works would have been as revelatory—and presumably incomprehensible—to most medieval scholars as Ben Franklin cracking open one of Einstein's books on relativity.

Although copies of Archimedes' mathematical treatises came to be available during the Middle Ages, not many scholars read them. Aristotle was ascendant at this time, his all-encompassing ideas sanctified by the Church. Archimedes, by comparison, was an intriguing, but nebulous, figure; many knew vaguely of his works, yet relatively few had actually slogged through them. It was one thing to utilize Archimedes' value for π, another to study how he arrived at the number.

The medieval buzz on Archimedes centered on his nonmathematical exploits. First, there was his mechanical ingenuity; every reader knew that he had single-handedly launched a ship from dry land, invented the irrigation screw, and constructed a working model of the cosmos. Then there was his legendary zeal for study, exemplified by the oft-told circumstance of his death—struck down by a Roman soldier while contemplating a geometric problem. And finally there was Archimedes, the brilliant military mind, whose defensive machines turned back Roman legions from the walls of Syracuse.

It took no mathematical training for a medieval humanist to read the biographical accounts by the fourteenth-century scholar Francesco Petrarca—Petrarch—who brought to the fore Archimedes' military achievements (including the false assertion that Archimedes had invented gunpowder and the cannon). Vitruvius, Livy, Pliny, and Plutarch also found readers among Europe's educated elite seeking to learn more about Archimedes. Where Archimedes' accomplishments are mentioned, they are restricted to his machines and his dedication to study; not a single mathematical treatise is named or described. The medieval Archimedes is a heroic figure, a designer of clever devices, an icon of scientific purpose—not yet the greatest mathematical mind of antiquity.

Petrarch also popularized the account by the Roman statesman, orator, and philosopher Cicero of his search for Archimedes' tomb. Like his upper-class peers, Cicero was a fanatic student of all things Greek, knowledge of whose culture and scholarship was considered the hallmark of a refined Roman. In 75 B.C., while serving as quaestor (adminstrator) in Sicily, Cicero set out to find the unique monument that marked where Archimedes had been laid to rest 137 years earlier. He was stunned to learn that Syracusan officials did not know of their illustrious countryman, much less where he was interred.

Cicero led a party of citizens to the necropolis outside the city's Achradina gate and sighted through the bushes a column bearing the unmistakable form of a cylinder and a sphere. Half of the mathematical epigram was visible, the other half worn away. Cicero directed workers to clear out the underbrush and restore some semblance of dignity to Archimedes' grave site. Writing thirty years later, Cicero was still outraged: "Thus the most famous city of the Greek world, once even the most learned, would have been ignorant of the memorial of its most keen-witted citizen, had it not learned of it from a man of Arpinum." Medieval intellectuals, only now acquainting themselves with classical forebears like Archimedes, no doubt took Cicero's millennium-old admonishment to heart. (Postclassical searches for Archimedes' tomb have turned up nothing, although one eighteenth-century traveler was helpfully shown the spot where Archimedes' house once stood and also the tower where he unleashed his burning mirror on the Roman fleet.)

The science historian Herbert Butterfield dubs the medieval Archimedes "the patron saint of the mechanically-minded." Given his reputation as a practical genius, Archimedes' name was used even into the Renaissance as an honorific for those who excelled in design or construction. The architect Filippo Brunelleschi was popularly known as the "second Archimedes." Leonardo da Vinci named his proposed steam-powered cannon "Architronito" in honor of the ancient sage.

To medieval humanists, Archimedes was a clever, practical man, but hardly to be counted among the giants of antiquity, such as Aristotle, Plato, and Euclid. As late as 1557, one Italian mathematician's ranking of classical "wise men," which placed Archimedes at the head of the list, drew this intemperate response from the humanist Giulio Cesare Scaliger: "From your own brain you hang the iniquitous scales, and weigh the worth of the ancient wise . . . You have put a builder [Archimedes] before Aristotle, who was no less knowledgeable in those arts . . . After Archimedes you have put Euclid, as if the light after the lantern. You seem to have been seized by the whirlwind and tempest of your own genius."

The transformation of the medieval Archimedes as designer-inventor into the Renaissance Archimedes as mathematician paralleled the burgeoning movement in Europe to recover and study classical Greek literature. These Greek-inspired stirrings would bloom into a full-fledged humanist Renaissance, surpassing Europe's twelfth-century, Arabic-based reawakening. Scholastic centers sprang up in many cities and, with them, humanist libraries. The owners of these mostly private libraries were collectors or scholars, and frequently both. Their motivation often rose above mere pride of ownership, resting on a forthright desire to restore humanity's lost intellectual heritage and thereby advance contemporary studies. Although not technically trained, they did purchase works in astronomy and mathematics, such as Ptolemy's astronomical tome, the *Almagest*, and, of course, Euclid's *Elements*. Contemporary correspondence reveals that, for the most part, owners of the major humanist libraries made Greek works available to scholars for translation into Latin. The circulation of Latin copies of library manuscripts proved essential to the advancement of medieval mathematics, whose practitioners would otherwise have found it nearly impossible to consult the treatises of their Greek predecessors.

The 1400s also saw the advent of a consultant class, educated men who applied mathematics and technical know-how toward the solution of practical problems: the design of buildings and public works; the development of military weapons and defensive battlements; the translation or editing of technical treatises from antiquity. With lucrative contracts available, instruction in the works of Euclid, Archimedes, and other classical mathematicians became the entry ticket for aspiring engineers during the Middle Ages.

Simultaneously, the dissemination of Plutarch's moralistic tales of famous ancient Greeks and Romans accelerated the shift in attitude toward Archimedes. This amended Archimedes disdains invention and engineering. His vaunted military work is but a distraction from his pursuit of mathematics for its own sake. As the linchpin of his argument, Plutarch points out that Archimedes published not a single treatise about his practical work. The reason is clear to Plutarch: Archimedes wished to distance himself from such base activities. The sixteenth-century essayist Michel de Montaigne paints a Plutarchian portrait of the beleaguered researcher that persists in popular accounts to this day: "The tale of that geometrician of Syracuse who was interrupted in his contemplations in order to put some of them into practical use in the defense of his country: he set about at once producing frightful inventions, surpassing human belief; yet he himself despised the work of his hands, thinking that he had compromised the dignity of his art, of which his inventions were but apprentice toys." With growing availability and understanding of his mathematical works, this latter-day glorification of Archimedes—martyred in the name of geometry—provided Renaissance scientists with a noble model.

In 1423, the scholastic centers of Italy were abuzz with the rumor that the antiquarian Rinuccio da Castiglione, in Bologna, had returned from Byzantium with a lost treatise of Archimedes, purportedly of a military nature. There was a furious exchange of letters among Italy's humanists to find out more—especially, the manuscript's whereabouts. To every inquiry, Rinuccio responded that he had lent it out, but never revealed to whom. In July 1424, in Florence, Friar Ambrogio Traversari confronted Rinuccio, who was passing through the city on his way to Rome. Rinuccio claimed to have the Archimedean manuscript in his luggage and promised Traversari a glimpse of the work the following day. Instead, Rinuccio left Florence and gave Traversari a wide berth on his return trip. In the end, no one ever claimed to have seen the purported Archimedean manuscript except Rinuccio himself.

Among the many caught up in the Rinuccio affair was Tommaso Parentucelli, then secretary to Bologna's Bishop Niccolo Albergati and a fervent bibliophile. Parentucelli ascended to the papacy in 1447 and, as Pope Nicholas V, immediately established a department within the Vatican to obtain and translate Greek classics. The "humanist pope," as Nicholas became known, dispatched agents to monasteries throughout his domain to acquire neglected works. His aim was no less than to create within the Church the greatest library since that of Alexandria—a goal aided by the fall of Constantinople to the Ottomans in 1453. By the time Nicholas died two years later, the Vatican manuscript collection surpassed all others in Europe, numbering 1,209 codices, of which 414 were in Greek.

Perhaps primed by the Rinuccio affair, it's no surprise that when a *real* Archimedean manuscript—Codex A—resurfaced around 1450, Pope Nicholas borrowed it for a new illuminated Latin translation by the scholar Jacobus Cremonensis. Who owned

Codex A remains a mystery. The Cremonensis translation now joined William of Moerbeke's from 1269 as a critical fount of Archimedean scholarship for European intellectuals. It lacks only Archimedes' key scientific work, *On Floating Bodies*. Scholars who wished to read this treatise had to consult William's Latin translation. If they wanted to read the work in the original Greek, they were out of luck. With the demise of Codex B, the sole extant Greek version of *On Floating Bodies* lay in the Archimedes Palimpsest, whose existence was unknown at the time.

In the early 1460s, Cardinal Bessarion, an influential émigré from the Eastern Church, borrowed the Vatican's Cremonensis translation to produce a copy for his own academy of the humanities. Bessarion's library became the largest treasury of classical Greek works in Europe, surpassing even the Vatican's. Apparently, Bessarion never returned the Cremonensis manuscript, for both it and its copy are in Bessarion's collection at the Marciana Library in Venice.

Bessarion subsequently hosted the gifted German astronomer-mathematician Regiomontanus, who produced his own corrected version of Bessarion's copy. Regiomontanus remarks in his correspondence that he had on hand an older Archimedes compilation, presumably William of Moerbeke's translation or perhaps the famous Codex A itself. In 1472, after returning to his native Germany, Regiomontanus seized upon Johann Gutenberg's introduction of movable type to become Europe's first publisher of printed books in science. But his plan to publish the works of Archimedes never materialized.

In 1491, Codex A was purchased by the prominent Venetian humanist Giorgio Valla, who reluctantly allowed it to be copied for the library of the Medici family. (This copy is currently housed in the Laurentian Library in Florence.) Upon Valla's death in 1500, eight hundred gold pieces brought Leo the Mathematician's famed codex to the statesman-scholar Alberto Pio, prince of Carpi, in northern Italy.

Pio permitted a copy of Codex A in 1544 for the royal library of King Francis I at Fontainebleau. The preface to the Fontainebleau manuscript notes that Codex A was so worn by this time that the name Archimedes could no longer be read on the title page. Scribes also complained of missing pages and obliterated text. Also in 1544, Thomas Gechauff Venatorius in Basel, Switzerland, published the first printed edition of Archimedes' known works, combining a copy of the Greek Codex A with the Latin translation by Regiomontanus. Another authoritative version in Latin was published in Venice in 1558 by Federigo Commandino, whose mathematical facility allowed him to complete several Archimedean proofs that had survived only in fragmentary form. Other printings followed, in Paris, Nuremberg (in German), London, and Oxford. No longer did one have to be wealthy or well connected to have access to the mathematical methods of Archimedes.

The Archimedean revival gathered momentum during the second half of the sixteenth century, and with it further evolution in the characterization of the man. Plutarch's prevailing vision of Archimedes as the exemplar of the pure mathematician conflicted with the impression conveyed by his treatises. What Plutarch had failed to recognize (perhaps willfully) is the occasional give-and-take in Archimedean reasoning between pure mathematics and the physical realm: the application of mathematical analysis toward real-world problems and, in the countersense, the examination of concrete forms to uncover fundamental geometric truths. Where Plutarch saw mathematics, he understandably dubbed the author a "pure" mathematician and played down the "baser" accomplishments. The truth is far richer: Archimedes' analytical language, ancient though it may be, is the precursor of the technical style increasingly adopted by physicists after the Renaissance. By the nineteenth century, a physicist without facility in mathematics was hardly considered a physicist by his colleagues. While Archimedes' works lack the theoretical underpinnings of modern-era science— the knowledge of forces, atoms, energy, and the like—his analytical form is evocative of the way modern-era physicists express their ideas.

This mode of scientific exploration, especially, with its emphasis on the application of mathematical analysis to physical problems and its occasional suggestion of measurement-based observation, offered Renaissance-era researchers—Galileo notable among them—an effective model with which to study nature. Galileo bypassed the seemingly ad hoc arguments of Aristotle for the rigorous, mathematical stylings of Archimedes. "The authority of Archimedes alone will satisfy everybody," states Galileo's mouthpiece Salviati in the 1638 essay *Dialogues Concerning Two New Sciences*. The historian Herbert Butterfield echoes Galileo's sentiments: "Archimedes . . . appears to have done something to assist and encourage this habit of mind and nothing could have been more important than the growing tendency to geometrise and mathematise a problem. Nothing is more effective, after people have long been debating and wrangling and churning the air, than the appearance of a person who draws a line on the blackboard, which with the help of a little geometry solves the whole problem in an instant."

Butterfield may be overstating the case for Archimedes' material impact on Renaissance science. To a degree, Archimedes did provide concrete results and effective working methods that were valuable to latter-day researchers. The young Galileo used Archimedes' center-of-gravity concept in a 1587 paper to secure his first appointment as a mathematics professor; a senior scholar who read the paper was mystified by the notion. Other mathematicians reintroduced Archimedes' take on the mathematical method of exhaustion. Without doubt, Archimedes was effective as a scientific muse, someone gifted almost beyond comprehension, whose intellect more properly placed him in an era far later than his own. Ask a gathering of Renaissance scientists to name the predecessor they would most eagerly transport into their midst, and they would overstep their medieval brethren in favor of Archimedes. To Galileo, Kepler, Boyle, Newton, and other Renaissance inheritors of Archimedes' legacy, their classical forebear was, in Galileo's words, nothing less than *divinissimus Archimedes*— the most divine.

Alberto Pio, prince of Carpi, died in 1550. He bequeathed his prized book, the faded, crumbling Archimedean Codex A, to his nephew, Cardinal Rodolfo Pio, like his uncle an avid collector of antiquities. When Rodolfo died in 1564, Codex A was not among the eighteen hundred manuscripts that passed from his estate to Modena's Estense Library. Nor has it been seen since. With the disappearance of Codex A, two of the three Byzantine compilations of Archimedes had run their respective journeys from creation to oblivion, having nonetheless bridged the centuries-long divide between the world of Leo the Mathematician and that of his Renaissance successors. Meanwhile, far beyond the ken of European minds, a forgotten fount of Archimedean wisdom, Codex C—the Archimedes Palimpsest—had already left its birthplace in Constantinople for a most unlikely destination: the Judean wilderness.

Chapter 12

GENTLEMAN AND SCOUNDREL

MS Additional 1879.23 Palimpsest.
Subscript: fragment of a mathematical work, with diagrams.
10th cent.
Vellum, 192 × 147 mm., 1 f.
Tischendorf, 1876.
—CAMBRIDGE UNIVERSITY LIBRARY CATALOG
ENTRY FOR AN UNIDENTIFIED MANUSCRIPT
PAGE, LATER DETERMINED TO BE FROM
THE ARCHIMEDES PALIMPSEST

THE MAR SABA Monastery, six hundred feet above the Kedron River, west of the Dead Sea, is a forbidding oasis of stone within an even grimmer landscape. The monastery's rough-hewn ramparts and towers hug the steep contours of the valley as though disgorged by the cliff face itself. Underneath Mar Saba's fortresslike structures lie a multitude of natural recesses once occupied by the valley's previous inhabitants: hermits and the dead. At its zenith, some ten thousand refugees had fled the commotion of the outside world for the seclusion of this oxymoronic community of hermits.

Founded in the fifth century by St. Saba, the originator of Orthodox monasticism and once a hermit himself, Mar Saba is among the oldest continuously inhabited monasteries in the world. St. Saba's preserved remains lie within the compound's domed, cruciform church, the Katholikon. Not far away, a chapel adjoins the cave St. Saba occupied as a hermit; here are displayed the skulls of monks slaughtered by Persian invaders in A.D. 614. One nineteenth-century visitor remarked that the "whole scene presents a confusion of small courts, chapels, churches, cells, projecting windows or terraces, and microscopic gardens, for every spot that will hold soil is utilised to redeem the savagery of the landscape by refreshing green."

The quiet that pervades Mar Saba today, with its handful of monks, belies its glorious past. In the seventh century, Mar Saba was a bustling intellectual and spiritual community of some four thousand residents. Among the tumble of buildings was a scriptorium for copying and illustrating manuscripts. And in the rectangular, fortified tower provided by Emperor Justinian to protect the compound was a library containing thousands of Greek manuscripts.

From an ex-libris once folded into its cover, we know that by the sixteenth century the Archimedes Palimpsest had joined Mar Saba's extensive manuscript collection. Who authorized the palimpsest's transfer from its birthplace in Constantinople, and why, nobody knows. Whatever the reason, it had nothing to do with Archimedes or his work. The intellectual climate in Byzantium for the study of mathematics had become as parched as the Mar Saba desert. To the Orthodox monks who bore the palimpsest from Constantinople along the rugged Kedron Valley to Justinian's tower at Mar Saba, the manuscript was a utilitarian liturgical document—a Euchologion—its underlying mathematical text and diagrams a shadowy remnant of a forgotten world.

The Euchologion would have been consulted regularly by Mar Saba's residents, both in their daily routine and for special occasions. In the tradition of the monastery's founder, St. Saba, the monks practiced the healing arts and would have availed themselves of the book's various rites of purification and exorcism. Judging from the candle-wax drippings on its pages, the Euchologion's utility continued well after sundown. Where gaps existed in its liturgy, monks inserted prayers and instructions written on paper sheets.

As centuries passed, Mar Saba's intellectual star guttered. Its once-treasured manuscripts became parchment stepchildren to monastic doctrine and ritual. Monks here and elsewhere in the barren deserts of the Middle East were inclined to use "surplus" manuscript pages for fuel. Surely something as obscure as a mathematical treatise would have been consigned to the flames without hesitation. But the eminently practical Euchologion was among the codices that were spared—and with it, quite inadvertently, Archimedes' writings.

By the mid-nineteenth century, history's inexorable tide swept the Archimedes Palimpsest to its next safe harbor. The Greek patriarchate, which oversaw Palestine's Orthodox Christian holy sites and relics, grew alarmed at the neglect and outright destruction of manuscripts in its far-flung monasteries. To better preserve its written heritage and make it more readily accessible to scholars, the patriarchate began a gradual centralization of its vast collection.

Around 1840, a series of manuscripts left Justinian's tower at Mar Saba for the patriarchate library in the Christian quarter of Old Jerusalem. In this shipment, still cloaked and forgotten beneath the Euchologian, was the Archimedes Palimpsest.

In short order, the restless palimpsest traveled again, first nearby to Jerusalem's famed Church of the Holy Sepulchre, and then, for reasons unknown, back to the city of its creation almost a millennium earlier, Constantinople. There it was added to the library of the Metochion ("daughter-house") of the Church of the Holy Sepulchre. And there, in September 1844, the palimpsest crossed paths with the noted—and notorious—biblical scholar from Leipzig, Constantin Tischendorf.

The early 1800s was a time of foment in the Christian community, when academic theologians engaged in fierce debate about the authenticity of the New Testament. Was the New Testament written by the disciples of Jesus or their near-contemporaries, or was it an ecclesiastical chimera—untrue and unholy—reflecting the religious whims of later clerics? The renowned British historian Edward Gibbon became a lightning rod for criticism when he asserted in his *Decline and Fall of the Roman Empire* that early Christian clerics had forged biblical documents in their effort to prosecute heretics. These documents, Gibbon wrote, were incorporated into subsequent renderings of the New Testament. Even supposedly authentic Greek Bibles might be tainted by such fictitious testimony. The debate spawned a new generation of manuscript hunters, who trekked in number to the paleological frontier—the old libraries of the Middle East.

Having made his startling accusations, yet unable to prove them, Gibbon withdrew from the ensuing debate. Gibbon's doubts about orthodoxy resonated with the Cambridge-educated scholar Richard Porson, who was on the fast track to an academic career before he derailed himself through drunkenness, irascibility, and his refusal to be ordained. But Porson's inebriation and ill-nature did not hinder his analytical skills. Faced with the challenge of determining which biblical documents were authentic—ones that reflected the actual text of Jesus's disciples—and which had been corrupted by later writers, Porson assembled a comprehensive list of commonalities among various texts: shared misspellings, alterations, transcription errors, or stylistic traits of a later age. These manuscripts, Porson concluded, are imperfect copies of earlier ones and do not reflect Holy Scripture. By the early 1800s, radical and conservative biblical scholars had reached an impasse on the authenticity of the New Testament. To the conservatives, one thing was clear: God's word was under assault.

Stepping foursquare into the midst of this theological whirlwind was Lobegott Friedrich Constantin Tischendorf, an energetic biblical scholar trained at the University of Leipzig. Tischendorf made his name in 1843 with his decipherment of the biblical palimpsest known as the Codex Ephraemi, which had previously defied transcription. Although a biblical conservative, Tischendorf agreed with the radical camp that ancient accounts of the Apostles had been seeded with errors—some willful, some mundane transcription flubs. Rather than evaluate the veracity of dubious biblical texts, Tischendorf proposed an alternative plan: He would scour libraries, imperial collections, holy sites, and monasteries of the Middle East to find the oldest, most pristine biblical documents and thereby "reconstruct if possible the exact text of the Bible as it came from the pen of the sacred writers."

In 1844, the twenty-six-year-old Tischendorf set off on the first of three biblical expeditions to the Middle East. He made brief stops in Malta, Alexandria, and Cairo before journeying into the Sinai wilderness to the Monastery of St. Catherine. Founded during the sixth century by Emperor Justinian on the purported site of Moses's burning bush, St. Catherine's possessed a manuscript library with fifteen hundred volumes in Greek and another seven hundred in Arabic. Resident monks insisted that there were no texts of any value, and collectors routinely left empty-handed. But Constantin Tischendorf was not so easily deterred. "Oh, these monks!" he fumed to his fiancée, Angelika. "I should be doing a good deed if I threw this rabble over the walls." Tischendorf was outraged at the monks' apathy toward their own manuscript collection. No discernible scholarship took place at St. Catherine's, and the monks were downright inhospitable to researchers from the Continent. His path to scriptural truth was being barred by religious cretins.

While poring through manuscripts in the monastery's library, Tischendorf spied parchment leaves from a fourth-century Greek Bible in a trash heap slated for burning. By his own account, the monks permitted him to take 43 of the 129 discarded sheets. (Given his grandiose sense of mission, Tischendorf's narratives of events must be viewed critically. To wit, the 43 leaves he took from St. Catherine's are described by him as deteriorated; in fact, they are in fine condition.)

Tischendorf would return to St. Catherine's in 1853 and 1859 to retrieve the Bible's remaining pages—"the pearl of all my researches," he called them—which together form the Codex Sinaiticus, the oldest-known complete New Testament in Greek. Having assured the monks in writing that he wished only to borrow the leaves to study and transcribe, Tischendorf instead presented them as a gift to his benefactor, Czar Alexander II of Russia. (Alexander sent the monastery nine thousand rubles in compensation, after which Bishop Callistratus of Sinai relinquished all rights to the codex. In 1933, the Russian government sold the Codex Sinaiticus to the British Museum, where it now resides.)

From the Sinai, Tischendorf traveled to monasteries in Jerusalem, Bethlehem, Mar Saba, and Constantinople. "Constantinople is a wondrous work," he enthuses in his travelogue, "heaven, earth, and sea created it, and art lent thereto a happy aiding hand . . . It is as if all the splendours of the world had concentrated themselves to discourse to the eye . . . To exaggerate a description of Constantinople would be very difficult, at least for the western European. Any picture of it, whether by the pencil or the pen, could only be an approximation to the beauties of the original."

Like his European contemporaries, Tischendorf's Ottoman reveries featured hidden storerooms bulging with Christian treasures and walled-up libraries full of manuscripts from antiquity. In September 1844, Tischendorf introduced himself to the former Greek Orthodox patriarch Constantius, who whetted his paleological appetite with the tale of the discovery in 1680 of a golden case containing the hand of John the Baptist. Foremost on Tischendorf's itinerary in Constantinople was the "old Patriarch's Library" at the Metochion of the Church of the Holy Sepulchre. Assuring himself of Tischendorf's bona fides, the Metochion's bishop permitted him free run of the repository.

Of the thirty liturgical manuscripts Tischendorf examined at the Metochion, only one captured his interest: a Euchologion with a faded technical text underneath. He offhandedly describes the document in his travel memoir as "a palimpsest dealing with mathematics." This brief remark is the first modern-era reference to the Archimedes Palimpsest. To Tischendorf, who was not mathematically trained, the technical symbolism was meaningless scrawl, not rare Archimedean writings. Whatever it was, it had nothing to do with the New Testament. Nevertheless, Tischendorf must have sensed the document's potential importance. He took out a blade, surreptitiously excised a sample page featuring both text and diagrams, and spirited it out of the Metochion.

That Tischendorf had violated the trust of the patriarchate was not revealed until thirty years later, when Cambridge University purchased the unidentified palimpsest page from his estate—along with several dozen other manuscript leaves Tischendorf had apparently pilfered during his lifetime. The severed palimpsest page became Cambridge Manuscript 1879.23. Hiding in plain sight on a library file card was the key to the existence of the Archimedes Palimpsest. Not until 1971 did Oxford's Nigel Wilson identify the source of the page: the palimpsest's text of Archimedes' treatise *On the Sphere and Cylinder*.

Driven by the wealth of liturgical treasures in the patriarchate libraries—and perhaps by the rapacious tendencies of scholars like Constantin Tischendorf—the Greek Orthodox patriarch Nicodemus, in the 1880s, commissioned a detailed catalog of the patriarchate's manuscripts. The Byzantine expert Athanasios Papadopoulos-Kerameus took nearly a decade to complete the four-volume tome: 2,556 pages describing some 2,350 Greek manuscripts at the various branches of the patriarchal libraries throughout the Middle East. Practically lost among the flood of entries in the catalog's fourth volume is manuscript number 355, a codex at the Metochion of the Church of the Holy Sepulchre in Constantinople: a Euchologion, circa twelfth century, palimpsest with unidentified mathematical text. Supplementing the Euchologion were a pair of sixteenth-century paper quires—signatures, in modern print-speak—inserted in different places among its pages. It was one of these quires that contained the now-lost ex-libris of Mar Saba, whose monks presumably authored the paper liturgies. Papadopoulos-Kerameus summarized the contents of the Euchologion for his catalog but, like Tischendorf fifty years before him, was stumped by the underlying mathematics. However, for researchers who might be so inclined, he included in his catalog description a sample of the peculiar writing.

While Papadopoulos-Kerameus was assembling his catalog, Johan Ludvig Heiberg, by then a renowned classicist at the University of Copenhagen, undertook his own Herculean task: to create the definitive Latin translation of all extant Archimedean treatises. Others had plowed the Archimedean fields, dating back to the thirteenth-century Greek-to-Latin translation by William of Moerbeke. But ancient Greek technical writing is tough to translate, especially when strewn with errors and abbreviations from serial transcribers. Heiberg believed that his predecessors misconstrued mathematical meanings and had a tin ear for Archimedes' Doric dialect. Having immersed himself for decades in the mathematics of the ancients, Heiberg was primed to introduce Archimedes to the twentieth century.

Through careful analysis, Heiberg deduced that, with a few minor exceptions, all extant copies of Archimedes' treatises formed an ancestral tree rooted in a common source: the pair of ninth-century Byzantine compilations he designated Codex A and Codex B. These ur-codices, dating to Leo the Mathematician, were long lost, forcing later scholars—Heiberg included—to rely on copies and derivative texts. These works were now scattered throughout the libraries and institutions of Europe. Other than fragments of treatises in Arabic, there simply was no other independent source of Archimedean treatises that Heiberg could consult. By the 1890s, the fragile cord that bears ancient writings forward through time had frayed into the slenderest of threads.

In the midst of his project, Heiberg received a note from a German colleague, Hermann Schöne, regarding manuscript number 355 in the new patriarchate catalog: a palimpsest whose Greek undertext was mathematical in nature. Heiberg studied the mathematical sample Schöne had copied out from the catalog entry. Although brief, the text's distinctive Doric inflection was unmistakable. Heiberg must have been stunned. Here was the voice of Archimedes calling out to him—but from a source he had never before encountered. Could there truly exist another Archimedean manuscript unknown to the world?

Through diplomatic channels, Heiberg petitioned the Turkish government to have the palimpsest sent to him for identification and study, a common courtesy within the academic circle. He was rebuffed. To read the manuscript, Heiberg would have to travel to Constantinople. By the summer of 1906, Heiberg was sitting in the library at the Metochion, magnifier in hand, paging through a previously unknown trove of Archimedean works, all in the original Greek. As he scanned the archaic writing, there was no doubt in his mind that the palimpsest's scholarly impact would be enormous. The style and format of the text indicated a very early origin, perhaps as long ago as the tenth century, fully three hundred years before William of Moerbeke's famous Latin translation. With its predecessor compilations A and B nowhere to be found, the palimpsest—dubbed Codex C by Heiberg—claimed the mantle as the oldest extant record of Archimedes' works. The news hit the front page of the *New York Times* on July 16, 1907: "Big Literary Find in Constantinople: Savant Discovers Books by Archimedes, Copied About 900 A.D."

The scholarly impact of the palimpsest was swift and profound. It contained complete or partial Greek texts of six known Archimedean works: *On the Equilibrium of Planes, On Floating Bodies, Measurement of a Circle, On Spirals, On the Sphere and Cylinder,* and the *Stomachion*. Into Heiberg's hands had fallen a wholly independent source for his own translation of Archimedes into Latin plus the means to reevaluate the various medieval and Renaissance versions. For four overlapping treatises, he could now contrast the palimpsest's Greek phraseology with that of an extant copy of the original Codex A, to better authenticate Archimedes' line of reasoning. The palimpsest also restored the original Greek text of *On Floating Bodies*, which had vanished with the demise of Codex B some six centuries earlier. Among his discoveries, Heiberg found an entire passage in William of Moerbeke's translation of the treatise that does not appear in the palimpsest; evidently, William drew on non-Archimedean sources where Codex B was illegible.

However, an even more profound gift awaited readers of the palimpsest. Heiberg's pulse must have raced when he encountered the title Εφοδος—*Method*—for this was a name he knew from tantalizing mentions in the tenth-century Byzantine encyclopedia *Suidas* and in the work *Metrica* by the first-century mathematician Heron. From these fragmentary clues, Heiberg had inferred the lost treatise's subject—and had lamented its disappearance. Now as the palimpsest's forgotten text seeped into his consciousness, he began to realize the significance of what lay open before him. "I deem it necessary to expound the method," Archimedes writes to his colleague Eratosthenes in Alexandria, "partly because I have already spoken of it and I do not want to be thought to have uttered vain words, but equally because I am persuaded it will be of no little service to mathematics." The great sage was about to explain to Eratosthenes—and now to Johan Heiberg—how he developed his geometric ideas.

Heiberg knew that classical Greek geometers routinely swept away all evidence of their working methods, leaving behind only polished, unerringly forward-flowing scientific proofs. Any hint of how they conceived the published steps—what convinced them to turn right or left at every logical fork—is absent. Archimedes was no different—except, Heiberg found to his delight, here in the pages of this tattered little book. Archimedes reveals that his geometric ideas didn't pop randomly into his head; nor did he uncover geometric relationships by trial and error. To the contrary, he employed a method. As Heiberg read on, he became convinced: This was Archimedes' lost treatise, the *Method of Mechanical Theorems*. Within the palimpsest's yellowed leaves, he had stumbled upon a map of Archimedes' mathematical mind.

Consider a loaf of sliced bread. Together all the slices preserve the overall form of the loaf. Measurement of any individual slice reveals geometric properties, such as height and width, at a certain location along the length of the loaf. Suppose the loaf, say, bulges at the middle; that feature will show up as a variation in a tally of the dimensions of the constituent slices. If the slices are thick, and therefore few in number, their array of measured dimensions will conjure an image that is only a jagged approximation of the loaf's true shape. The thinner the slices, the more of them there are, and the more closely their array of dimensions hews to the actual form of the loaf. In principle, an infinite number of slices, each infinitesimally thin, would mirror the loaf's shape precisely; that is, this array of measurements alone is sufficient to recreate the loaf. While impossible to implement in the real world, an infinite number of slices is feasible in the conceptual realm of mathematics. The analytical process that applies to a loaf of bread applies as well to a cylinder, sphere, cone, or any other geometric form. Once such a form is rendered into its numerical avatar, true mathematical analysis can begin.

Archimedes' self-styled mechanical method entails the figurative slicing of various shapes as the first step toward computing their surface areas and volumes. However, his key innovation in the *Method* is fortifying this mathematical procedure with concepts from the branch of real-world physics known as mechanics—the study of the stability and movement of physical bodies. Specifically, Archimedes brings to bear his own law of the lever in what would ordinarily be a purely abstract exercise. In his imagination, he suspends various pairs of object-slices from the ends of a lever and explores the conditions that bring the lever into balance. The result of this geometric manipulation is a quantitative measure of the volume of the entire object before it was sliced.

Among the *Method*'s fifteen propositions are detailed reckonings of the center of gravity of spheres, hemispheres, and rounded cones. Other propositions compare the area or volume of various pairs of geometric figures: a section of a parabola versus an inscribed triangle; a sphere versus an inscribed cone; a cylinder versus an inscribed rectangular prism; and, Archimedes' personal favorite (engraved on his tombstone), a cylinder versus an inscribed sphere.

By way of example, in proposition 4, Archimedes explains how he applies his mechanical method to express the volume of a

paraboloid—a shape similar to the rounded nose cone of a rocket—
in terms of the volume of a circumscribed cylinder. Archimedes
chooses a cylinder as the comparative benchmark because he had
previously determined the formula for a cylinder's volume. He as-
sumes that these geometric forms are real, that they have heft, so
when they are suspended from the ends of a lever, the lever will tilt
or balance under their combined weight. Although he doesn't say,
Archimedes might have even fashioned a paraboloid and a cylinder
out of wood or metal and observed the effect of their placement
on the tilt of a lever.

Archimedes was unable to draw firm mathematical conclusions
by considering the paraboloid and the cylinder in their entirety.
He therefore analyzes them slice by slice. Figure 12-1 shows the
initial arrangement, with the nested paraboloid and cylinder affixed to the right half of a lever, line *HD*, like marshmallows on a
stick. The fulcrum (not shown) is situated midway along the lever
at point *A*. That is, *HA = AD*. The midpoint of the cylinder—its
center of gravity—is located at point *K* on the lever; the cylinder

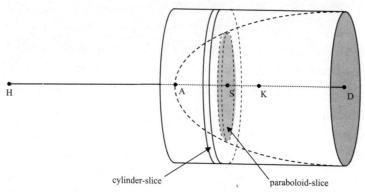

Figure *12-1. Archimedes' method applied to a paraboloid and circumscribed cylinder*

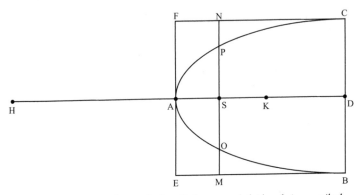

Figure 12-2. Archimedes' method applied to a paraboloid and circumscribed cylinder, here depicted in cross-section

bears downward on the right half of the lever as though all of its weight were concentrated at *K*. Figure 12-2 shows the same arrangement in cross-section. In this view, the paraboloid appears as a two-dimensional curve—a parabola—threading points *C*, *A*, and *B*. The circumscribed cylinder appears as a rectangle bounded by points *C*, *B*, *E*, and *F*.

Now the *Method* takes an unexpected turn. Archimedes excises a thin, circular slice of the paraboloid between points *P* and *O* and suspends it from the far end of the lever at point *H*. (In the cross-sectional view, the circular slice appears simply as a line.) The corresponding slice of the cylinder, the circle between points *N* and *M*, remains where it is. Archimedes demonstrates that the paraboloid-slice suspended at point *H* balances the cylinder-slice in its original location around point *S*. Successive slices are made. Each time, the paraboloid-slice is added to the previous slices suspended at point *H* and the corresponding cylinder-slice remains where it is. In the end, the entire mass of the paraboloid has been amalgamated at point *H*, while the intact cylinder remains centered around point *K*. Archimedes has effectively transformed the original arrangement of shapes into a straightforward lever problem involving a pair of weights, one at H and one at K, with the fulcrum in between at A. Now he needs only to observe how far point H must lie from the fulcrum to bring the lever into balance.

The ratio of the lever arms HA and AK reveals the relative weights—and therefore relative volumes—of the paraboloid and the cylinder. The heavier, and more voluminous, shape must sit closer to the fulcrum. From figure 12-2, it is evident that the lever arm of the cylinder, AK, is half that of the lever arm of the sliced-and-reassembled paraboloid, HA. Thus Archimedes arrives at his goal: The volume enclosed by a paraboloid is precisely half that of the circumscribed cylinder.

Of the method, Archimedes reminds his correspondent Eratosthenes that "the fact here stated is not actually demonstrated by the argument used; but that argument has given a sort of indication that the conclusion is true." In other words, Archimedes admits that his method is an end run around the rigorous Euclidean proof of a geometric assertion. And Archimedes did provide Eratosthenes with the requisite hard proofs of the *Method*'s various propositions. But, he adds, "it is of course easier, when we have previously acquired, by the method, some knowledge of the questions, to supply the proof than it is to find it without any previous knowledge." Although Archimedes conceived his method only as a practical guidepost to geometric truths, the procedure would pass muster today as an acceptable form of proof in mathematics.

Heiberg might have doubted his own eyes when he perused particular passages in the *Method*. The underlying concept was familiar to him—slicing geometric forms into exceedingly thin elements—but the time period was all wrong. This was a foundational idea of

calculus, a branch of computational mathematics developed by
Newton and Leibniz during the seventeenth century. True, nowhere
in the writing that Heiberg perused does Archimedes specify that
he slices his forms into an *infinite* number of elements, as is the case
in calculus, but to Heiberg the implication was inescapable. Heiberg
also realized that Archimedes' union of pure mathematics and principles of physics overleaps the ages as well. As with calculus, the
potent analytical tools of mathematical physics began to emerge
during the seventeenth century. With the *Method of Mechanical
Theorems*, Archimedes anticipated by some two thousand years
Europe's most fertile mathematical minds. Of course, these latter-
day thinkers were unaware of the treatise's existence. The *Method*
lay forgotten among the pages of the palimpsest, far removed
from the traveled byways of Western science and mathematics.
Renaissance researchers knew of Archimedes only through his
other works, which together were sufficiently far-reaching to cast
him as a genius. Had the *Method* been resurrected earlier, their admiration for Archimedes would have been all the greater and, perhaps, speeded the development of modern mathematics.

In 1907, Heiberg published the Greek text of the *Method*, and
from this, Thomas Heath produced an English translation in 1912.
Between 1910 and 1915, Heiberg published his revised collection
of Archimedes' works in three volumes, this edition informed by
his close reading of the palimpsest as well as the discovery of the
Method. Heiberg's side-by-side Greek-Latin renderings became the
standard reference work for Archimedean scholars.

The palimpsest emerged intact from World War I to witness
Turkey's declaration as a sovereign nation in January 1921. Later
that year, Greek forces swept into Turkey to topple the government
of Mustafa Kemal—Atatürk. Fearing for the safety of the Meto-

chion's irreplaceable manuscripts, the patriarch of the Greek Orthodox Church secretly ordered the transfer of the monastery's
890 texts to the National Library of Greece. Only 823 arrived.
The Archimedes Palimpsest was not among them. The winds of
chance, stirred by human acquisitiveness, had wafted the fragile
codex from Constantinople all the way to Paris.

Chapter 13

THE FRENCH CONNECTION

The earliest extant manuscript of the works of Archimedes, the unique source for his Method of Mechanical Theorems, *the only surviving witness to the original Greek text of* On Floating Bodies, *the most substantial and significant Greek palimpsest known, arguably the most important scientific codex ever offered at auction.*

—FROM *THE ARCHIMEDES PALIMPSEST*, CHRISTIE'S AUCTION CATALOG, 1998

DESPITE HIS BEST efforts, Johan Heiberg had been unable to wrest the Archimedes Palimpsest from the grip of Turkish government functionaries back in 1906. Antiquities were not to be removed from Turkish soil, they had declared, no matter how learned the petitioner or compelling the project. If Heiberg wished to peruse Archimedean treatises that had lain dormant for seven hundred years, he had to do so not in far-off Copenhagen but sitting at a desk in the library of the Metochion in Constantinople. Faced with this breach of scholarly etiquette,

Heiberg had settled for a photographic copy of the codex and a brief reinspection of the original two years later. That the Archimedes Palimpsest did leave Turkey not long afterward reveals the essential flaw in Heiberg's strategy: He had asked permission. The paper trail of the Archimedes Palimpsest during much of the twentieth century is meager, and whatever evidence exists is shaky. Altogether, witnesses' attestations paint a crude, often contradictory picture, blurred by poor recollection and colored by self-interest. Who possessed the Archimedes Palimpsest up to 1908, when Heiberg last saw it, is clear: the Metochion—and by extension, its nominal supervisory body, the Greek Orthodox patriarchate. Also evident is that sometime after 1908, the palimpsest left the Metochion and passed through any number of hands until it settled in a private collection in Paris no later than about 1930. As though awakening from some decades-long slumber, the palimpsest had reemerged into the realm of the verifiable. Who possessed the palimpsest from this time until it burst onto the public scene in the late 1990s has been established with reasonable certainty. Who, in fact, was its rightful owner was settled only after a dramatic legal wrangle in 1998.

Unlike the text of the palimpsest, so ably transcribed for modern minds, the late history of the palimpsest may never be fully extricated from the tangled testimony that surrounds it. The true account of the book's latter-day existence is itself a palimpsest, the factual record sponged away by time and overwritten by a host of competing narratives. Yet in the spirit of the palimpsest's ardent champion Johan Heiberg—who was, at base, a history detective—let's parse these metaphorical strata of evidence to deduce how the case of the missing palimpsest might have unfolded. And yet, when all done, we must screw up our eyes again to read the most recent layer of this palimpsested tale—one still in process—that lays out a compelling alternate history.

After completing his military service in Greece during World War I, the French businessman and civil servant Marie Louis Sirieix returned to the region in search of antiquities. Around 1921, Sirieix acquired the Archimedes Palimpsest from a dealer in Constantinople. How the palimpsest had come to be removed from the library of the Metochion is unknown. Other codices had been sold by Metochion personnel as early as 1905, joining rare book repositories at Duke University, the University of Chicago, the Cleveland Museum of Art, the Walters Art Museum in Baltimore, and the Bibliothèque nationale de France in Paris. Evidently, the patriarchate exerted no oversight of the care or disposition of its manuscripts. By 1921, the palimpsest's unique legacy was either forgotten or unappreciated by both its Greek Orthodox stewards and its new owner.

In the tumultuous environment of 1920s Turkey, Sirieix easily spirited the palimpsest out of the country to his apartment in Paris at 54 Rue de Bourgogne near the Musée Rodin. To its new owner, who was unschooled in antiquities, the manuscript was a Byzantine religious book—a Euchologion. It was ancient, to be sure, but otherwise unremarkable. Not the kind of book, he was told, that would quicken the pulse of an antiquarian collector. The underlying mathematical text was nothing more than a curiosity to Sirieix. The most desirable books of this type, he came to find out, were embellished not with geometric diagrams but with ornate, multicolored illuminations, some trimmed in gold leaf. The deficiency of his Euchologion must have gnawed at Sirieix, for sometime after 1938, he appears to have made a radical decision. As one might slap fender skirts and whitewalls on an old convertible, Sirieix evidently contracted with an artist—well, to be specific, a forger—to enhance the sale potential of his homely codex.

Like his predecessor in the year 1229, the forger saw in the Euchologion-Archimedean codex little more than a ready source of parchment. He first removed its sixteenth-century binding. (Indications are that the Mar Saba monks had rebound the codex around that time following a fire.) He then excised four leaves from various parts of the work—pages that held passages from the Archimedean treatises *On Floating Bodies*, the *Method*, and *On the Equilibrium of Planes*. Like the original Euchologion scribe some seven hundred years before, he washed the leaves in an attempt to create a blank canvas, but the acidic oak-gall ink had etched itself permanently into the parchment.

After the severed leaves had dried, the forger opened his copy of Henri Omont's *Miniatures des plus anciens manuscrits grec de la Bibliothèque Nationale du VIe au XIVe siècle*, published in 1929. From plate 84 of Omont's book, the forger transferred onto the four Euchologion leaves meticulous tracings of the evangelists, images Omont had photographed from an illuminated eleventh-century Byzantine gospel text. Luke was drawn over an Archimedean passage from *Floating Bodies*, John and Matthew over different parts of the *Method*, and Mark over *Equilibrium of Planes*. Perhaps pressed for time, the forger ignored the complicated background scenes in the original illustrations and substituted a generic Byzantine-style arch to frame each figure. Guessing the original hues (Omont's plates were in black-and-white), he painted the fake evangelists in multicolor, brushed on a thin gilding suspension, and, when done, gouged the pictures with a knife to give them an authentic distressed look. Flecks of gilded Archimedean treatises littered the tabletop.

To augment his "improvement" of the Euchologion, the forger copied onto another page a small decorative element from Omont's book. In a final—and apparently unique—sprucing up of the work, either the forger or a successor glued to the inside of the Euchologion's cover a pair of carefully trimmed canon tables, cut from a thirteenth-century Byzantine gospel book. (Neither the Metochion's cataloger Papadopoulos-Kerameus nor Johan Heiberg mentions the presence of such tables when they saw the codex.)

This wasn't the only time the forger consulted the photographs in Omont's *Miniatures*. A twelfth-century Byzantine gospel book (now owned by Duke University) contains identical portraits and decorative elements painted over preexisting text. That book, like the Archimedes Palimpsest, had been cast out of the Metochion's collection. The claim by some paleographers that the evangelist portraits in the Euchologion-Archimedean codex were produced not by a Parisian forger but by the Metochion monks themselves before Sirieix acquired the manuscript is no longer tenable, given both the 1929 release date of Omont's book and the recent identification of a pigment in the paint that was not available in Europe until 1938.

Overall the Euchologion's fake illuminations would have looked compelling to a neophyte collector; however, in their distinctly non-Byzantine hues and abridged composition, an expert would have suspected a fraud. In fact, the entire enterprise was misguided. As time would judge, the monetary and scholarly value of the codex lay in its obliterated Archimedean writings. Marie Louis Sirieix (or whoever hired the forger) had tarted up the Euchologion unnecessarily. As if mocking its owner, peeking through the diaphanous gilding above Luke's haloed head, like a mischievous cartoon thought-bubble, are several lines of the codex's text.

Along the edges of the evangelist portraits, the rusty stigmata of paper clips suggest that these pages had a life of their own separate from the palimpsest. At intervals, they must have been clipped to cards for display. Evidently, paper clips proved insufficient to the task; traces of adhesive, including one formulation unavailable in Europe until 1970, are found on the backs of the illustrations. Each portrait is numbered in pencil, 1 through 4, in the upper-right corner. There may have been up to seven forged illustrations altogether covering pages from the palimpsest. The present-day codex is three leaves short compared to Johan Heiberg's count from 1906. Minute traces of paint speckle the facing pages where these lost leaves had once been. Somewhere in a museum or a private collection might be three Omont-sourced, Byzantine-style evangelist portraits masking leaves from the Archimedes Palimpsest.

In 1947, Sirieix moved from Paris to the south of France, leaving his apartment and the Archimedes Palimpsest to his daughter Anne Guersan. (Sirieix died in 1956.) Guersan, working with her husband and son, began to explore the possibility of selling the palimpsest around 1970. Initially, the Guersans appear to have had no idea of either its scholarly significance or its value. All they knew for sure was that the codex had deteriorated from years of improper storage, its pages now splotchy with mold. The decrepit little book looked ready for the trash. (Heiberg's 1906 photos of the palimpsest show no such deterioration; apparently, the damage occurred on Sirieix's watch.)

To assist in the conservation and eventual sale of the manuscript, the Guersans showed several leaves to Professor Abraham

Wasserstein, an expert on classical Greek mathematics at Jerusalem's Hebrew University. In a letter to the family dated August 25, 1970, Wasserstein confirms that the leaves belong to the Metochion's manuscript 355 and quotes from its 1899 catalog entry by Papadopoulos-Kerameus. Wasserstein mentions the manuscript's mathematical text, but nowhere does he identify the timeworn book with Heiberg's famous Archimedean codex. Given the appalling condition of the sample leaves, Wasserstein recommends that the Guersans have the manuscript restored at an establishment in Paris.

The Guersan family also showed the manuscript to their upstairs neighbor, Jean Bollack, a noted classics professor at the University of Lille. Decades later, Bollack recalled informing the Guersans at the time of the codex's exceptional provenance and, like Wasserstein, advised them to have the mold treated. The date of this verbal exchange is uncertain. Bollack claims it took place as early as 1960, although if true, the Guersans' 1970 consultation with Wasserstein appears redundant—the family would have already learned the significance of their mold-ridden heirloom. Or perhaps, given the potential worth of the palimpsest, they were seeking a confirmatory opinion.

Also around 1970, the Guersans brought sample leaves from the palimpsest to the Institut de Recherche et d'Histoire des Textes in Paris. Here the manuscript was definitively identified by the antiquarian book expert Joseph Paramelle. (Curiously, no record exists of the palimpsest's several-week stay at the institute.) The Guersans subsequently followed the consultants' advice and delivered the entire palimpsest to a private establishment in Paris for conservation.

Starting shortly thereafter, the Guersans distributed two hundred copies of a brochure titled "Archimedes and Codex C or Eureka"—half of the copies in English, the other half in French—offering the palimpsest for sale. By this time, only part of the palimpsest remained bound into its current nineteenth-century sheepskin-over-wood cover; the rest, including the four evangelist portraits, were loose sheets. Inquiries came in from a number of institutions, including Yale, the University of Texas, the University of Pittsburgh, and the antiquarian book dealer H. P. Kraus in New York. Although the Guersan family kept up its marketing efforts into the 1980s, no sale resulted. Potential buyers were likely put off by the family's high asking price and the manuscript's less than secure provenance. As with other codices removed from the Metochion, the Greek Orthodox patriarchate did nothing at this time to dispute the Guersans' claim of ownership of the palimpsest.

In 1993, having failed in its own efforts, the Guersan family consigned the sale of the Archimedes Palimpsest to Christie's auction house. At first, the French Ministry of Culture denied the family's request for an export permit, only to reverse itself upon appeal in 1996. The palimpsest began its passage to the United States via Oxford University. Here it underwent study by the noted classics scholar Nigel Wilson, who subsequently provided the technical narrative for Christie's tony auction brochure. (Wilson had previously identified the Tischendorf leaf at Cambridge as having been stolen from the palimpsest.) The Archimedes Palimpsest arrived at Christie's in New York in 1998. The auction was announced for the afternoon of October 29. On October 28 at 5:00 P.M., the patriarchate filed for a temporary restraining order to block the sale, asserting that the palimpsest had been stolen from the Metochion.

In a parallel effort to bypass the auction, the Greek minister of culture declared his government's desire to purchase the Archimedes Palimpsest, casting its plan as a historical, moral, and scientific imperative. The manuscript, according to the minister, formed an essential part of the Greek cultural legacy. (This despite the fact that when the Metochion's manuscripts were secretly transferred for safekeeping to the Greek National Library in the 1930s, no inventory was taken to ascertain the presence of the palimpsest.) Before auction day, a consortium of Greek institutions offered the Guersans $1 million for the palimpsest. The offer was declined.

On October 29, 1998, promptly at 2:00 P.M., an announcement was read by a Christie's representative to the gathered audience: "Christie's is pleased to inform its clients that the Federal Court in New York last night denied a motion by the Greek Orthodox Patriarchate of Jerusalem to enjoin this afternoon's sale of the Archimedes Palimpsest. The judge ruled that under the applicable law our consignor has clear title to sell the manuscript, and the sale will take place as scheduled." With that, an image of a page from the codex flashed onto the screen at the front of the room. Bidding began on lot 1, the Archimedes Palimpsest.

Among the bidders that day were the Greek consul-general, acting on behalf of his government; several other members of the audience, along with half a dozen remote clients routed through a phone bank of Christie's representatives; and an anonymous private collector represented by his on-site intermediary, the prominent London antiquities dealer Simon Finch. (The collector later revealed that he was present at the auction.) Princeton University, which had declared its intention to participate in the auction, had withdrawn at the request of the Greek government.

Bidding started at $480,000 and rapidly escalated in $50,000 increments to $1 million. From there, the price rose with each bid by $100,000. The Greek government folded at $1.9 million. When the gavel came down, not five minutes into the auction, the Archimedes Palimpsest had sold to Simon Finch's anonymous client for $2 million. The house premium upped the total purchase price to just over $2.2 million. The identity of the purchaser remains secret, although Finch allowed that he is a private American involved in the high-tech industry—specifically, *not* Bill Gates. (The German magazine *Der Spiegel* reports that the palimpsest's owner is most likely Jeff Bezos, founder and CEO of the online retailer Amazon.com.)

Following the auction, the patriarchate sued Christie's, Anne Guersan, and the anonymous buyer for the return of the Archimedes Palimpsest. The patriarchate asserted that Guersan had no right to sell the palimpsest in the first place, since no one at the Metochion had authority to dispose of it.

First the critical matter of legal jurisdiction had to be settled. Whose laws governed the establishment of ownership of the palimpsest? The patriarchate petitioned New York's Southern District Court that the case be decided under New York law, which has strong protections for the rights of original owners of disputed works of art. Framing its argument somewhat tangentially in terms of commercial considerations rather than legal principles, the patriarchate contended that New York City's reputation as a preeminent cultural and marketing center for art would suffer if the sale of stolen goods was sanctioned. Clearly, New York had a stake in the outcome of the suit. Guersan herself benefited from the city's high standing among art patrons. Therefore, the merits of the case should be adjudicated according to New York law.

However, Judge Kimba Wood sided instead with the defendants, who argued that French law applied. That the palimpsest was auctioned in New York, the defense asserted, was immaterial. All that mattered was where the *original* transfer of title to the property took place. Based on the various affidavits, the palimpsest was presumed to have passed to Guersan's father, Marie Louis Sirieix, either in Constantinople or in Paris. Whichever, Sirieix certainly did not purchase the manuscript in New York. Nor did Judge Wood buy the patriarchate's argument regarding the auction's deleterious impact on art commerce in New York City.

Under French law, good-faith purchasers of allegedly stolen works assume ownership after thirty years of "continuous and uninterrupted, peaceful, public and unequivocal possession." Therefore, the case hinged on two factors that overrode the seemingly insoluble question of whether the palimpsest was stolen: whether the Guersan family possessed the palimpsest for the prescribed duration and in the prescribed manner; and whether the patriarchate made a timely and significant effort to recover its property.

Did the Guersan family possess the palimpsest for at least the requisite thirty years? The patriarchate's lawyer pressed Anne Guersan to produce a bill of sale or similar document confirming her father's transaction. However, Judge Wood ruled that it was unreasonable to expect the defendant to produce a bill of sale after so many decades. Had the patriarchate acted sooner than seventy years after the fact, such a document might have been available. It would be unfair to disadvantage Guersan's case because of the patriarchate's failure to act promptly.

While a valid bill of sale would have settled the ownership issue, the Guersans did not have to prove possession all the way back to the 1920s, but only back to 1968, thirty years before the lawsuit. Their only documentation was the present-day testimony of Professor Bollack that he had been shown the palimpsest around 1960. Judge Wood ruled that although Bollack's stated date of 1960 was uncertain, he was unlikely to have erred by a decade.

As to the public aspect of possession prescribed by French law, here, too, Judge Wood favored the arguments of the defense. The Guersan family had not hidden their possession of the Archimedes Palimpsest. Indeed, they had shown the manuscript to various experts during the 1970s, if not earlier; had announced its availability for purchase through brochures mailed to major institutions; and had negotiated with potential buyers, including one face-to-face meeting in their home. Judge Wood concluded that the Guersan family had met all the requirements of French law to legitimately claim ownership of the palimpsest.

The patriarchate's case was weak on several points. It produced no evidence that the palimpsest had been stolen from the Metochion. Nor did it dispute the fact that the Metochion had sold similar works, all of which are openly held by major institutions. Prior to bringing its lawsuit in 1998, the patriarchate made no effort to locate or retrieve the palimpsest. In truth, it appears to have acted only in the wake of negative publicity in Greece. Its claim that, as a religious entity, it lacked the disposition to mount a timely search for its property found no sympathy with the court. Judge Wood shot back that "if the Patriarchate was able to retain counsel with impressive speed to bring this action the night before the Christie's auction, it could have retained counsel to search for the Palimpsest, or at least make some inquiries, at some point during the previous seventy years." The decades-long delay in pursuing its claim clearly prejudiced the defendants' case; where documentation might still exist, where memories might still be fresh, where crucial parties like Guersan's father, Marie Sirieix, might still be alive—all were lost as a result of the patriarchate's inaction. In short, so many years had elapsed that evidence-based judgment was impossible.

On August 19, 1999, Judge Wood dismissed the case against Christie's, Anne Guersan, and the anonymous owner of the Archimedes Palimpsest. The ancient codex, having spent centuries in obscurity before bursting into the spotlight, officially passed to the next station in its long existence. And yet, from the historian's perspective, nothing had been settled in court but matters legal and monetary. What about the historical facets of the case? The most important mathematical document of antiquity disappears in the early twentieth century, then reappears decades later. The dubious accounting of its whereabouts in those intervening years is as patchy as the palimpsest itself. Nevertheless, for lack of any concrete evidence, the tale recounted in Judge Kimba Wood's courtroom would have to suffice, perhaps for the ages, as a placeholder for an unknowable truth.

The magnitude and seeming insolubility of this mystery weighs on the historian. To be able to set down in such rich detail all that has been learned about the long-ago Archimedes and his works, and yet to be unable to peer into the darkness of decades so proximate to our own is colossally frustrating. But if history holds any lesson for its practitioners, it is this: Every now and then, when the historian is about to walk away and pursue a more promising avenue of study, serendipity casts its adventitious beam into the gloom and illuminates a new—perhaps wholly different— historical truth. In the case of the Archimedes Palimpsest, that metaphorical beam manifested itself in the form of the real-life scholar Georgi Parpulov, who, in 2006, uncovered a seven-decades-old letter by the palimpsest's true owner, a man whose name never came up in Kimba Wood's courtroom. In that forgotten letter lies a second lesson for the historian: Solving one mystery inevitably breeds another.

On February 10, 1934, a Saturday, the Parisian antiquities dealer Salomon Guerson sat down to compose a letter to the distinguished New Testament scholar Harold R. Willoughby at the University of Chicago. Now sixty years old, the Turkish-born Guerson was frustrated. For too long he had been trying to sell his old Byzantine codex, but no one was willing to meet his price. True, the book was smallish and had evident signs of age. And, yes, it lacked the colorful illuminations that made collectors' eyes light up. Still, Guerson believed that Professor Willoughby would be sympathetic to his plight. When they met in late 1931, the professor had given him reason to believe that his book was worth a considerable amount of money. Perhaps Willoughby's University of Chicago—or even Willoughby himself—would be interested in purchasing the manuscript. Nevertheless, it had been two years since they had corresponded; before contemplating such an outlay, the professor might need to be refreshed on the unique nature of the artifact in question. Business is business, no more so than in antiquities, and the terrain of negotiation must be laid.

Born in Turkey in 1873, Salomon Guerson was a veteran of the Byzantine antiquities trade. As a young man in the 1890s, he had established an office in Constantinople—the city the Turks of late preferred to call Istanbul. There he had conducted a lucrative business alongside fellow antiquarians like his friend Divran Kelekian, who became a top-tier dealer in rare books. Guerson had gravitated to textiles—antique carpets and tapestries—but kept his eyes open to opportunities in parchment as well. He had long ago left Constantinople to set up shop on Boulevard Haussmann in Paris. In rare textiles, Guerson was expert; in rare books, not quite so. Nevertheless, he had surmised that the book he was trying to sell was unique. And that is why he had enlisted Professor Willoughby's help.

Guerson began his current letter to Willoughby by recounting their meeting two years ago. He had carried his precious codex across the Atlantic aboard the *Olympic* as it steamed from Cherbourg to New York in November 1931. Like many other old parchment texts Willoughby had encountered, this one was a palimpsest. However, the undertext was unlike any Willoughby had run across. It was mathematical in nature and accompanied by fine-lined geometric diagrams. As Willoughby must have explained to Guerson at the time, his specialty was New Testament iconography, not Greek mathematics. They needed the input of an expert to evaluate and identify the manuscript. Guerson agreed and, following an introduction by Willoughby, dispatched one of his palimpsest's loose leaves to Reginald Berti Haselden, curator of manuscripts at the Huntington Library in Pasadena, California.

Haselden had been experimenting with ultraviolet photography in the decipherment of ancient manuscripts. He hoped that such nondestructive techniques might replace the devastating chemical interventions that had been used to render a palimpsest's lower text visible. Amazingly, from the single leaf he had been given, Haselden confirmed the extraordinary nature of Salomon Guerson's book. The underlying Greek text on the parchment was identical to a passage on page 248 of Johan Heiberg's momentous 1907 article in the journal *Hermes*. Without doubt, the leaf was folio number 57 of Heiberg's Codex C. Salomon Guerson had in his possession the celebrated Archimedes Palimpsest.

With the preliminaries out of the way, Guerson now got to the point of his letter to Willoughby: The Archimedes Palimpsest was still for sale. He had already approached Henri Omont, curator at the Bibliothèque nationale, who had recently published that informative book of photographs of Byzantine religious art. But Omont was unwilling to meet Guerson's asking price for a manuscript— even the vaunted Archimedes Palimpsest—that lacked illustrations. The Bodleian Library at Oxford had turned down the palimpsest as well. "You would greatly oblige me," Guerson pitched, "if you would let me know whether this manuscript interests you or at any rate if you would write to me to whom I could offer it with a chance of selling it. I am asking $6,000." In today's currency, Guerson's price was a steep $98,000.

At this point, the twentieth-century chapter of the Archimedes Palimpsest again trails off into obscurity. Willoughby's reply to Guerson's 1934 letter is lost, although it's clear that the palimpsest did not wind up in Chicago or, it appears, anyplace else for at least the next several years. However, by means unknown, the palimpsest emerged from World War II in the hands of a different owner, the Parisian government worker Marie Louis Sirieix. And this postwar palimpsest contained forged paintings of the evangelists on four of its pages, including the text leaf analyzed by Reginald Berti Haselden in 1932.

Who ordered the forgeries, Guerson or Sirieix? Of the two, Guerson was surely more knowledgeable about Byzantine art. His recently discovered letter to Harold Willoughby confirms that he knew Henri Omont and presumably Omont's book of photographs from which the evangelist paintings were copied. Furthermore, at the Paris exhibition of Byzantine art in 1931, Guerson contributed a strikingly similar Omont-sourced forgery on a leaf detached from a different Metochion book. The Archimedes Palimpsest was on the market for an extended period during the 1930s, unsold, and Guerson may have grown desperate as the political situation in Europe deteriorated. As a Jew, his position in Paris became increasingly precarious, especially after the Germans arrived in 1940. The evidence implicates Salomon Guerson as the person who arranged the defacement of the palimpsest, perhaps as an eleventh-hour effort to make it more salable. The palimpsest's subsequent owner, Marie Sirieix, was more likely responsible for the appalling condition of the manuscript after World War II. Salomon Guerson might have grudgingly consented to painting over Archimedes' words, but, as a longtime collector, he surely would have avoided damaging his manuscript through improper storage.

The palimpsest that reemerged so publicly in 1998 looked different from the one pictured in Johan Heiberg's photographs. More damage had been done to it during the twentieth century alone than during the previous eight. Two million dollars bought a moldy carcass of a manuscript, defaced as much through neglect as through willful action. After its brief reappearance, the Archimedes Palimpsest might have retreated again into obscurity. But rather than secrete away the unique document, the palimpsest's anonymous owner turned it over to researchers in the hope of extracting further insights into the mind of Archimedes. Nearly a century after Johan Heiberg perused the manuscript by lamplight in the Metochion, scientists directed a host of modern technologies toward the oldest extant record of Archimedes' achievements. It soon became clear that the palimpsest had yet more secrets to reveal.

Chapter 14

SWEETEST SUSTENANCE OF SOULS

The whole earth is the tomb of heroic men and their
story is not given only on stone over their clay but
abides everywhere without visible symbol woven into
the stuff of other men's lives.

—PERICLES

VERY, VERY, VERY ugly" was the reaction of William
Noel, curator of manuscripts and rare books at the Walters
Art Museum in Baltimore, after he withdrew the Archimedes
Palimpsest from the duffel bag on his desk. "It doesn't look like a
great object at all." As he and the museum's conservator of manuscripts, Abigail Quandt, paged through the codex, the damage
wrought by recent decades of neglect sank in. Like a nightmarish
botanical version of flesh-eating bacteria, mold had overrun the
palimpsest so severely in places as to have eaten through the
parchment. Some leaves were so fragile they threatened to crum-
ble at a touch. Nor had intact areas been spared, now stained a
sickening purplish color by the moldy infestation. And, of course,

there were the four evangelist portraits, slathered as appallingly over Archimedes' writings as whitewash over a Rembrandt.

Even as the patriarchate pressed its legal case for the return of the Archimedes Palimpsest in 1998, William Noel pursued a parallel mission: to borrow the palimpsest for exhibition and study. Noel contacted the palimpsest's anonymous owner through his representative, Simon Finch. The reply was both prompt and surprising. Not only did the owner consent to lend the palimpsest to the Walters Museum, known for its conservation of historical artifacts; he also agreed to fund the first modern-era research of the aged manuscript. On the specified day, the owner himself brought the costly palimpsest to Noel's office at the Walters—in the duffel bag.

Before the research could begin in earnest, Abigail Quandt had to see to its protection and conservation. Having removed the palimpsest's cover, she attempted to free the manuscript from its sewn binding that researchers might glimpse text formerly hidden within the book's spine. (Heiberg had not been allowed to take apart the codex, so spine-bound sections of Archimedes' text remained out of sight.) Quandt was dismayed to find that many of the quires had been not only sewn together but glued.

Binding books with adhesive is a relatively recent practice. The adhesive applied to the palimpsest was PVA—polyvinyl acetate, or hardware-store-variety wood glue—a formulation developed only after World War II. Evidently, the palimpsest's previous owner, the Guersan family, tried to secure loose quires that were falling out of the spine. The intervention worked too well; the PVA glue bonded with the parchment and was now almost impossible to remove without taking with it pieces of parchment. Aided by a microscope, Abigail Quandt patiently brushed a solution of iso-propynol and water onto pieces of adhesive, then tweezed the softened flakes off the fragile matrix of the parchment. The entire process took four years, the final glue fragment removed in late 2004. Simultaneously, Quandt scraped off the innumerable beads of wax that had drizzled from monks' candles over the centuries, flattened leaves that had been creased or folded, and reinforced fragmented sections with lightweight Japanese paper. With con-servation complete, the palimpsest was handed over to scientists.

A variety of techniques have been applied to improve the read-ability of the Archimedean text beyond what Heiberg had been able to eyeball. After World War I, ultraviolet light was introduced to help readers discern the lower script of palimpsests. These invis-ible rays are the synthetic version of the component of sunlight that causes exposed skin to tan or burn. With its relatively high en-ergy, ultraviolet light penetrates and activates materials more effec-tively than visible light.

The efficacy of ultraviolet imaging of old manuscripts rests on the presence of iron-containing pigments in the ink. By absorbing the ultraviolet illumination, the iron-rich ink fluoresces against its more muted parchment backdrop. Photographs of this differential glow delineate the shapes of the characters that constitute the text. The process effectively dials up the contrast between the writing and the parchment.

That the chemical composition of the Euchologion ink differs somewhat from that of the underlying Archimedean ink boosts the effectiveness of the imaging of the palimpsest: By adjusting the wavelengths of light used for illumination and detection, re-searchers have generated photographs in which the Euchologion text virtually disappears. Nevertheless, the easiest-to-read images have proven to be those in which the Euchologion and Archimedean texts are both present but artificially rendered in contrasting colors. Imaging by a consortium of researchers from the Rochester Institute of Technology, Boeing, and Equipoise Imaging has revealed Archimedean text even on pages where none is visible to the eye. (A digital version of the Archimedes Palimpsest is available on the Internet at www.archimedespalimpsest.org.)

Despite their high energy, the ultraviolet rays were unable to penetrate the obscuring paint of the four evangelist forgeries. The failure of the ultraviolet imaging technique left scholars with a thorny dilemma. Should these nonhistorical paintings be chemically or mechanically removed to reveal the hidden work of Archimedes? Or should they be left in place, an unfortunate but essential element of the palimpsest's tangled history and a cautionary example of the desecration of our cultural heritage?

The dilemma has largely been rendered moot by technology: Where ultraviolet rays failed, X-rays have succeeded. X-rays are far more energetic than ultraviolet rays and exhibit sufficient penetrating power to breach a wide array of visibly opaque materials—including a forger's paint. Following feasibility studies with a tabletop X-ray device, full-scale X-ray fluorescence imaging was conducted in 2005 at the Department of Energy's Stanford Synchrotron Radiation Lightsource (SSRL). Here electrons are accelerated to nearly the speed of light within a 260-foot-diameter circular pipe, emitting a hair's-width X-ray beam that is directed onto a target. As a precaution, researchers test-fired the SSRL's beam at a sample of parchment before exposing the palimpsest to the barrage of energy.

The SSRL's targets are ordinarily much smaller than a page-size piece of parchment. To expose every part of the page to the

X-rays, researchers developed a motor-driven mounting frame that moved the page incrementally across the stationary beam. At each spot, the beam passed through the evangelist portrait and caused the underlying ink to fluoresce in a way that could be recorded by a detector. Like a pointillist painting, an overall image of the hidden writing was assembled from the multitude of individual measurements.

With dental or medical X-rays, a film plate or digital sensor array is placed behind the target, creating a ghostlike negative image; however, such a conventional arrangement isn't sensitive enough to portray the subtle contrast between ink and parchment. At the SSRL, the detector was situated off at an angle in front of the page. The image derived not from the primary X-ray beam itself but from secondary X-rays that originated in the fluorescing ink. Fluorescence X-rays are sensitive to the elemental composition of materials and therefore served to highlight the contrast between ink and parchment and even between different formulations of ink. It was through X-ray imaging that the name of the Eucholo-gion scribe and the date of the palimpsest's creation were discovered.

With the advent of computers, fluorescence images that have been digitized—rendered into arrays of numbers that gauge the intensity at various points of the image—can be mathematically processed to further enhance legibility. Similar computer-based algorithms have been applied to palimpsest pages that have been scanned by high-resolution cameras. Derek Walvoord at the Rochester Institute of Technology in Rochester, New York, has developed an algorithm to identify the many fragmentary Greek characters that are otherwise difficult to interpret. Acting on a digitized avatar of the palimpsest, this optical character recognition software effectively learns the Archimedean scribe's handwriting by analyzing the pen strokes that constitute the various characters. If successful, the process will reveal text on pages that have suffered the most severe degradation by mold.

What have Archimedes scholars gained from the application of modern technology? The text has been imaged more distinctly than what Johan Heiberg had been able to discern by eye, whether studying the palimpsest itself or its photographic copy. This is especially true for the *Method* and *Stomachion*, two works for which Heiberg had no prior model. As a result, hundreds of corrections have been made to Heiberg's version of the palimpsest treatises. Scholars have also filled in numerous gaps in Heiberg's transcription, places where he found the text to be either illegible or physically inaccessible, that is, hidden within the book's spine. In one case, an entire page of the treatise *On Floating Bodies* had been missed. (Heiberg photographed only pages on which he discerned erased text—just 103 of the palimpsest's 354 pages; the rest were ignored.)

Modern imaging has also revealed that the intellectual treasure of the palimpsest extends beyond the realm of Archimedes. Back in the thirteenth century, when the Byzantine scribe Ioannes Myronas prepared to pen the Euchologion, he needed more fresh parchment from his recycling heap than Archimedes alone could provide. There are at least five other Byzantine-era manuscripts buried under the Euchologion narrative: two complete speeches— ten full pages of text—by the famous Athenian orator Hyperides, a contemporary of Aristotle and Demosthenes; an anonymous commentary on Aristotle's treatise on categories; a liturgical work on the life of a saint; and a portion of an as-yet unidentified philosophical text.

One of the most significant interpretive opportunities arises from the palimpsest's resurrected Archimedean diagrams. In his publications, Heiberg replaced the palimpsest's diagrams with his own; his goal was not historical accuracy but stylistic uniformity and clarity of presentation for today's reader. Heiberg's line drawings follow twentieth-century conventions of diagrammatic representation; they are a literal visual translation of Archimedes' text. When a triangle is specified in the proof, a classic-looking triangle appears in Heiberg's diagram. When a pair of line segments are characterized as parallel, Heiberg depicts them tracking precisely alongside one another. None of this is surprising to the modern student of geometry, who has been taught to draw figures that hew precisely to the text. Sure, some mathematicians draw better than others, especially when using a computer-based drawing program, but purposeful artistic license is barred. By such a standard, Archimedes' diagrams appear almost playfully unconventional.

How confident are we that the diagrams in the palimpsest reflect Archimedes' own? If Heiberg introduced new diagrams in his modern transcription, might not his tenth-century forebear have done the same? Might the palimpsest's diagrams be the idiosyncratic work of an unknown Byzantine scribe—or any of his predecessor copyists—instead of faithful renderings of antiquity's greatest mathematician? The diagrams in the palimpsest are the end-product of serial copying, yet at most four or five generations removed from Archimedes' originals. Medieval copies of classical mathematical treatises reveal that diagrams in the exemplar were generally carried over to the new edition with minimal modification. Indeed the diagrams in the palimpsest's version of

Archimedes' treatise *On the Sphere and Cylinder* are virtually identical to those in extant copies of the long-lost Codex A. Scribes were copyists, not mathematicians. As savants of the written word, they were alert to misspellings and grammatical blunders. But the mysterious diagrams of classical geometers were beyond their ken. It was the rare scribe who possessed the intellectual means to evaluate such drawings. For the rest, the surest means to timely completion of their task lay in the line-for-line reproduction of the exemplar figures. Thus modern-day scholars are confident that the palimpsest's diagrams provide them with a new window into Archimedes' thinking.

If there is a latter-day successor to Johan Heiberg, it is Reviel Netz, professor of ancient science at Stanford University, who has studied the palimpsest's resurrected text and diagrams more than anyone else. Netz asserts that the drawings of classical Greek geometers are more than an illustrative aid to the written narrative; they form an integral part of the geometric proof, as the various sections of an orchestra combine to produce an overall sound. Greek treatises tend to leave out explanatory descriptors, such as *line* or *triangle*, words on which modern readers rely to guide themselves through a complex geometric narrative. Figures are often specified only by a sequence of letters; reference to a particular triangle might appear in the text as merely ΔΦΓ, without the prefatory *triangle*. Nor is there any mention in the text as to which pair of letters identifies the triangle's base and which the sides; this information came from the diagram. Netz has concluded that classical mathematicians expected readers to thread constantly between text and diagram. The illustrations in a geometric proof form a visual grammar that is key to its presentation.

When it came to creating diagrams, Archimedes was more

Pablo Picasso than Chuck Close. In a modern-era geometric or scientific argument, there is no discordance between the text explanation of a principle and its diagrammatic alter ego. They are mirror images of each other, only one is rendered in words and symbols, the other in lines, curves, and labels. But in the palimpsest, Netz sees Archimedes' figures as largely symbolic, where strictly defined Euclidean shapes are willfully stylized. There are many ways to represent, say, a polygon. Polygons have different numbers of sides, different side-lengths, and different internal angles. Some polygons are regular, others lopsided.

In Netz's view, Archimedes was concerned that when he added an illustration of an archetypal polygon to a geometric proof, the reader might erroneously link a *general* principle governing polygons to the pictured polygon alone. To allay such confusion, Archimedes needed to alert the reader that the polygon he drew was indeed generic. He could have done so explicitly within the text or by adding an overt disclaimer to the diagram. Instead, he signaled the figure's prototypical nature by visual means. In his treatise *On the Sphere and Cylinder*, for example, Archimedes replaces the stock polygon with a stylized one: a multisided figure whose boundaries are scalloped instead of straight. Similar diagrammatic alterations appear in the *Method*. Modern-day readers would be mystified by such odd riffs on standard shapes; classical mathematicians apparently expected them. To Archimedes' contemporaries, the altered shape was merely a visual cue that the figure on the page was to be considered generic, that it represented "polygon-ness" rather than any particular polygon. So what if it looked hokey.

Through new imaging techniques, Reviel Netz and his colleagues were also able to read critical passages in Archimedes'

Method of Mechanical Theorems that Heiberg had been unable to decipher. In the *Method*, Archimedes computes the volumes of an array of geometric forms through an innovative union of mathematics and physics: He slices comparative pairs of figures into thin elements and balances these slivers on a lever. However, the *Method*'s proposition 14, one of the narratives Heiberg found illegible, stands out from the rest. Here Archimedes casts aside his lever-balancing procedure and develops a wholly different means of volume calculation.

In proposition 14, Archimedes imagines a vertical cylinder. He makes an angle-cut through it, starting at the edge of the top face and carving uniformly down to the midline of the base. Removing the larger piece, the section of the cylinder that remains resembles an upright tapered fingernail, thick at its base, progressively thinner toward its upper end. It is this odd shape whose volume Archimedes computes. (The actual proof involves more general shapes, but this representative form is easier to picture.)

According to the new critical reading by Netz, Archimedes divides the figure into an infinite number of slices, then puts forward a computational technique to sum up the volumes of these slices directly. True, classical Greek mathematicians had envisioned an infinite set of numbers—no matter how large the quantity specified, there is always a quantity an increment larger—but in proposition 14, Archimedes goes a significant step further. He applies the notion of infinity not merely in its conceptual sense but in explicit service of calculation. He discusses the *plethos*, or numerical nature, of infinity. Just as he might observe that one triangle has the same number of corners as another—three—he equates, say, the infinite number of points that constitute a line to the infinite number of line segments that constitute a rectangle. The number 3 and the number *infinity* differ only as to their magnitude. Any doubt that Archimedes was plying the waters of seventeenth- and eighteenth-century mathematics more than two thousand years ago is laid to rest in these newly revealed passages from the *Method*.

From tenth-century urban Byzantium to the wilderness of Mar Saba; from the Church of the Holy Sepulchre in Jerusalem to the Metochion in Constantinople; from an apartment in Paris to an auction house in New York. From Constantin Tischendorf to Johan Ludvig Heiberg; Salomon Guerson to Marie Louis Sirieix; Anne Guersan to a nameless titan of the Internet age. From light to darkness, darkness to light; rumor to reality, obscurity to fame. For a thousand years, the Archimedes Palimpsest has ridden the roiling waves of circumstance to become the most celebrated link to antiquity's greatest mathematician-inventor.

This fragile emissary from the past has been scrubbed with acid, singed by fire, slathered with paint, sliced with a knife, and overrun by mold. Of late, it has been snipped, tweezed, and X-rayed in the name of science. In preserving the varied work of Archimedes, the palimpsest affirms that the best of ancient minds were every bit as capable as those of today. Granted, we stand on a higher rung in factual knowledge and physical understanding than our classical-age predecessors. But Archimedes' treatises reveal a degree of native reasoning ability and analytical thought that would pass muster in any era. Indeed Ben Jonson's homage to another acclaimed genius, Shakespeare, applies as aptly to Archimedes: "He was not of an age, but for all time."

In laying the foundations of science, Archimedes stands shoulder to shoulder with luminaries like Newton, Darwin, Maxwell, and Einstein. He exemplifies the human species' utmost reach as scientific thinker and mechanic, combining the footloose imagination of a da Vinci, the quantitative mastery of a Gauss, and the methodical inventiveness of an Edison. Archimedes could wrap his ken around a universe, then craft a whirring model of that cosmos on his tabletop; crank an imaginary pulley, then apply a real one to haul a ship into the sea. In turning abstract levers into marauding weapons, Archimedes demonstrated, too, the polar extremes of the scientific endeavor, from its purest to its most destructive. He was antiquity's one-man research-and-development institute—with the resources of an entire kingdom at his disposal.

Tallying Archimedes' accomplishments is easy, although it leaves one almost breathless with admiration. Pinning down Archimedes himself is harder. There was no ancient Boswell or Pepys to record his everyday life. Early chroniclers were more apt to elevate—or diminish—their subjects than provide factual grist for the mills of future historians. Even from the modern perspective, Archimedes' intellectual horsepower and evident eccentricities almost tempt one to ponder whether he was deposited here from another planet. No surprise that classical-era biographers viewed him as superhuman. No surprise either that Archimedes' true life and character remain entwined with the legendary. On a personal level, he continues to baffle us today as he surely did his fellow Syracusans. By what gauge do we situate him within the matrix of humanity when the available record presents him as unique? Nevertheless, a broad-brush portrait of Archimedes reveals a number of indisputable traits—creativity, productivity, inventiveness, mathematical and practical prowess—all of which seem to have burned brighter in him than in virtually anyone else throughout history.

Archimedes is gone, but his voice is preserved through his writings. Ideas, once recorded, are eternal, so long as there exist

inquisitive brains to receive them. Yet rising above even his manifold contributions to knowledge is Archimedes' enduring power to
inspire. As he nourished his own mind through constant intellectual
challenge, he set an example for generations of successors who likewise seek nature's fundamental truths. Cicero called this essentially
Archimedean pursuit the "sweetest sustenance of souls," and every
true scientific explorer is driven onward by its sublime energy.
Refracted through the life, legend, and legacy of Archimedes is the
timeless image of the scientist, patiently applying unbiased eye, hand,
and mind toward the explication of nature. And every once in a
while, if the experiment is apt, if the circumstances are right, if
the will bears up against repeated failure—

Eureka!

NOTES

Chapter 1: THE ESSENTIAL ARCHIMEDES

6   "From this day forth Archimedes is to be believed": Dijksterhuis (1987), 15.

9   "high spirit": Plutarch, *Marcellus*, in Stevenson (2000).

9   "sordid and ignoble": Ibid.

11 "included a hundred-and-eighty-foot golden phallus": Stille (2000), 90.

13 "continually bewitched by a Siren": Plutarch, *Marcellus*.

13 "possessed by a great ecstasy": Ibid.

13 "forget his food and neglect his person": Ibid.

14 "Archimedes possessed so high a spirit": Ibid.

Chapter 2: THE STORMY SEA

20 "Humans are beings who by nature": Aristotle, *Politics*, 1253a 2–3, in Martin (1996), 55.

21 "One adult son [from each family]": Crawford and Whitehead (1983), 59.

23 "And thus, the foundation is complete.": Ibid., 62.

23   "Every legislator ought to know": Plato, *Laws.*

23   "banish meanness and covetousness": Ibid.

25   "It is Sicily as a whole": Thucydides, *History*, book IV.59, in Crawford and Whitehead (1983), 407–8.

26   "an utter calamity": Ibid., Book VII.85, in Crawford and Whitehead (1983), 424.

27   "too cold for winter and too hot for summer": Cole (1844), 241.

29   "Where happiness derives from gorging": Benjamin (2006), 62.

30   "So many were employed in the teaching": Freeman (1891–94), vol. 3, 411.

31   "On Sicily's fields": Herman Melville, *Timoleon, Etc.* (New York: Caxton Press, 1891).

Chapter 3: EUCLIDEAN FANTASIES

35   "imam of mathematics": Berggren (1987), 101.

36   "is not possible to find in all geometry": Plutarch, *Marcellus.*

36   "One feature which will probably most impress": Heath (1912), vi.

37   "without exception, monuments of mathematical exposition": Heath (1981), 20.

37   "inventive to the point of playfulness": Netz (2004), 11.

38   "those who claim to discover everything": Heath (1897), 151.

44   "center is everywhere": *The Oxford Dictionary of Quotations*, 4th ed., ed Angela Partington (Oxford: Oxford University Press, 1992), 16, no. 17.

Chapter 4: NUMBER GAMES

53   "I will try to show you": Heath (1897), 221.

56   "There was a young fellow": George Gamow, *One Two Three . . . Infinity* (New York: Viking Press, 1961). In the original, the phrase "the square root of infinity" was expressed in mathematical symbols.

61   "these things will appear incredible": Dijksterhuis (1987), 373.

61   "A problem which Archimedes devised": Vardi (1998), 305.

62   "If thou art diligent and wise": Ibid.

63   "But come, understand also all these conditions": Ibid.

Chapter 5: EUREKA MAN

76   "This seems, so to say, a crude thing": *The Golden Crown: Galileo's Balance*, at Rorres Web site.

77   "Since the wires are very fine": Hoddeson (1972), 17.

Chapter 6: THE SCIENCE OF FEAR

87   "acquired the sovereignty of Syracuse": Polybius, *Histories*, book VII, part viii.

88   "The base of it was exactly square": Plutarch, *Demetrius*, in Stevenson (2000).

90   "despising his father as a dotard": Livy, *History of Rome from Its Foundation*, book XXIII, part xxx, at Rorres Web site.

90   "exceedingly capricious and violent": Polybius, *Histories*, book VII, part ii.

91   "Those who write narratives of particular events": Ibid.

93 "by natural inclinations addicted to war": Plutarch, *Marcellus*.

93 "did not reckon with the ability of Archimedes": Polybius, *Histories*, book VIII, part ii.

95 "There was discharged a piece of rock": Plutarch, *Marcellus*.

95 "a dreadful thing to behold": Ibid.

95 "the Romans, seeing that indefinite mischief": Ibid.

96 "When Archimedes began to ply his engines": Ibid.

96 "geometrical Briareus": Ibid.

97 "The Syracusans were but the body": Ibid.

Chapter 7: THE VOICE BENEATH THE PAGE

108 "the greatest literary loss the world ever suffered": *New York Times*, July 16, 1907.

108 "before the great day": Ibid.

Chapter 8: A BRIDGE ACROSS TIME

116 "My little fellow has opened his hand": Fischer (2001), 68.

117 "It is to writings that you must set your mind": Ibid., 45.

117 "The ear of a boy is on his back": Jackson (1981), 21.

119 "A big roll is a big nuisance": Seid (2004).

Chapter 9: THE PARCHMENT BROTHERS

122 "If you do not know what writing is": Jean (1992), 83.

127 "his eyes ran over the page": Martin (1994), 68.

128 "A cloister without bookcases": Banks (1989), 54.

129 "St. Patrick of Armagh, deliver me": Kilgour (1998), 71.

129 "Thank God, it will soon be dark": Avrin (1991), 224.

129   "Now I've written the whole thing": Ibid.

129   "For him that stealeth": Ibid.

130   "Neither my fame nor my praise": Jackson (1981), 70.

130   "Pleasant to me is the glittering": Ibid.

Chapter 10: LEO'S LIBRARY

132   "Apart from a handful of works": Browning (1964), 3.

136   "No person shall obtain a post": Wilson (1983), 2.

137   "O law, my brain could grasp": Treadgold (1979), 1260.

137   "What has Athens to do with Jerusalem?": Browning (1964), 3.

138   "Pagan culture was neither rejected nor accepted": Ibid., 4.

142   "dear friend of the Muses": Christie's (1998), 41.

Chapter 11: RESURRECTION AND LIGHT

146   "It has always been my opinion": Clagett (1982), 356.

147   "As a light shining in darkness": Ibid., 357–58.

148   "Here in the exemplar": Ibid., 365.

151   "Thus the most famous city of the Greek world": Jaeger (2002), 52.

151   "the patron saint of the mechanically-minded": Butterfield (1965), 104.

152   "From your own brain you hang the iniquitous scales": Laird (1991), 635–36.

153   "The tale of that geometrician of Syracuse": Simms (1995), 72.

157   "The authority of Archimedes alone": Galilei (2005), 194.

157   "Archimedes . . . appears to have done something": Butterfield (1965), 25–26.

Chapter 12: GENTLEMAN AND SCOUNDREL

161   "whole scene presents a confusion": Geikie (1887), chap. 31.

164   "reconstruct if possible the exact text": Bentley (1986), 44.

165   "Oh, these monks!": Ibid., 84–85.

165   "the pearl of all my researches": Tischendorf (1867), 23.

166   "Constantinople is a wondrous work": Tischendorf (1847),
          262; 267.

166   "a palimpsest dealing with mathematics": Christie's (1998),
          17.

170   "I deem it necessary to expound": Heath (1912), 13–14.

175   "the fact here stated": Ibid., 17.

175   "it is of course easier": Ibid., 13.

Chapter 13: THE FRENCH CONNECTION

186   "Christie's is pleased to inform": Rickey (1998).

188   "continuous and uninterrupted": Lowden (2003), 243.

189   "if the Patriarchate was able to retain counsel": Reyhan
          (2001), 1002.

193   "You would greatly oblige me": Netz and Noel (2007), 165.

Chapter 14: SWEETEST SUSTENANCE OF SOULS

196   "Very, very, very ugly": *Archimedes' Secret*, transcript of
          BBC television documentary, 2002.

208   "sweetest sustenance of souls": Cicero, *Tusculan Disputations*,
          in Jaeger (2002), 52.

BIBLIOGRAPHY

Africa, T. W. "Archimedes Through the Looking Glass." *Classical World* 68 (1975): 305–8.

Andrewes, Anthony. *The Greek Tyrants.* London: Hutchinson University Library, 1966.

Archibald, R. C. "The Cattle Problem of Archimedes." *American Mathematical Monthly* 25 (1918): 411–14.

Archimedes. *Geometrical Solutions Derived From Mechanics.* Translated by J. L. Heiberg and Lydia G. Robinson. Chicago: Open Court, 1909.

Assis, Andre Koch Torres. *Archimedes, the Center of Gravity, and the First Law of Mechanics.* Montreal: Aperion, 2008.

Avrin, Leila. *Scribes, Script and Books: The Book Arts from Antiquity to the Renaissance.* Chicago: American Library Association, 1991.

Banks, Doris H. *Medieval Manuscript Bookmaking: A Bibliographic Guide.* Metuchen, NJ: Scarecrow Press, 1989.

Bell, A. H. "The 'Cattle Problem.' By Archimedies (*sic*) 251 B. C." *American Mathematical Monthly* 2, no. 5 (May 1895): 140–41.

Benjamin, Sandra. *Sicily: Three Thousand Years of Human History.* Hanover, NH: Steerforth Press, 2006.

Bentley, James. *Secrets of Mount Sinai.* New York: Doubleday, 1986.

Berggren, J. L. "Archimedes Among the Ottomans." *Acta historica scientiarum naturalium et medicinalium* 39 (1987): 101–9.

———. "History of Greek Mathematics: A Survey of Recent Research." *Historia Mathematica* 11 (1984): 394–410.

Bergmann, Uwe. *X-Ray Fluorescence Imaging of the Archimedes Palimpsest: A Technical Summary.* http://www.slac.stanford.edu/gen/com/images/technical%20summary_final.pdf.

"Big Literary Find in Constantinople." *New York Times,* July 16, 1907.

Boas, Marie. *The Scientific Renaissance, 1450–1630.* London: Collins, 1962.

"The Booksellers of Antiquity: Authors and their Public in Ancient Times." Book review. *New York Times,* January 1, 1894.

Boyer, Carl B. *The History of the Calculus and its Conceptual Development.* New York: Dover, 1959.

———. "Quantitative Science without Measurement: The Physics of Aristotle and Archimedes." *Scientific Monthly* 60 (1945): 358–64.

Bragg, Melvyn. *On Giants' Shoulders.* New York: Wiley, 1999.

Browning, Robert. "Byzantine Scholarship." *Past and Present* 28 (July 1964): 3–20.

Bryden, D. J., and D. L. Simms. "Archimedes as an Advertising Symbol." *Technology and Culture* 34 (1993): 387–91.

Burns, Alfred. "Ancient Greek Water Supply and City Planning: A Study of Syracuse and Acragas." *Technology and Culture* 15 (1974): 389–412.

Bury, John B. *A History of the Eastern Roman Empire.* New York: Russell and Russell, 1965.

Butterfield, Herbert. *The Origins of Modern Science, 1300–1800.* New York: Free Press, 1965.

Campbell, Duncan. *Besieged: Siege Warfare in the Ancient World.* New York: Osprey, 2006.

Cartledge, Paul, ed. *Cambridge Illustrated History of Ancient Greece.* Cambridge: Cambridge University Press, 1998.

Carvalho, David N. *Forty Centuries of Ink.* http://www.world wideschool.org/library/books/tech/printing/FortyCenturiesof Ink/toc.html.

Caven, Brian. *Dionysius I: War-Lord of Sicily.* New Haven, CT: Yale University Press, 1990.

Christie's, Inc. *The Archimedes Palimpsest.* New York: Christie's, 1998.

Clagett, Marshall. "Archimedes." In *Dictionary of Scientific Biography*, edited by Charles C. Gillespie, 213–31. New York: Charles Scribner's Sons, 1970.

———. *Archimedes in the Middle Ages.* Philadelphia: American Philosophical Society, 1978.

———. *Greek Science in Antiquity.* New York: Barnes and Noble Books, 1994.

———. "How Archimedes Expected to Move the Earth." *Centaurus* 5 (1958): 278–82.

———. "The Impact of Archimedes on Medieval Science." *Isis* 50 (1959): 419–29.

———. "Leonardo da Vinci and the Medieval Archimedes." *Physis* 11 (1969): 100–151.

———. "William of Moerbeke: Translator of Archimedes." *Proceedings of the American Philosophical Society* 126 (1982): 356–66.

Cole, Thomas. "Sicilian Scenery and Antiquities." *Knickerbocker* 23 (March 1844): 236–44.

Crawford, Michael H., and David Whitehead. *Archaic and Classical Greece: A Selection of Ancient Sources in Translation.* Cambridge: Cambridge University Press, 1983.

Culham, Phyllis. "Plutarch on the Roman Siege of Syracuse: The Primacy of Science over Technology." In *Plutarco e le Scienze,* edited by Italo Gallo, 179–97. Genoa: Sagep Editrice, 1992.

Dehn, Max. "Mathematics, 300 B. C.–200 B. C." *American Mathematical Monthly* 51 (1944): 25–31.

Dickinson, G. Lowes. *The Greek View of Life.* Ann Arbor, MI: University of Michigan Press, 1958.

Dijksterhuis, E. J. *Archimedes.* Princeton, NJ: Princeton University Press, 1987.

D'Israeli, Isaac. *Curiosities of Literature.* 1793–1823. http://www.spamula.net/col.

Drachmann, A. G. "Archimedes and the Science of Physics." *Centaurus* 12 (1967–68): 1–11.

Drögemüller, Hans-Peter. *Syrakus: Zur Topographie und Geschichte einer griechischer Stadt.* Heidelberg: Carl Winter Universitätsverlag, 1969.

Easterling, P. E. "Hand-List of the Additional Greek Manuscripts in the University Library, Cambridge." *Scriptorium* 16 (1992): 302–23.

Eddington, Arthur Stanley. *The Nature of the Physical World.* New York: Macmillan, 1929.

Edwards, I. E. S., John Boardman, John B. Bury, and S. A. Cook. *The Cambridge Ancient History.* Cambridge: Cambridge University Press, 1969.

Fabricius, Knud. "Das Antike Syrakus." *Klio* 28 (1932): 1–30.

Finley, Moses I., Dennis Mack Smith, and Christopher Duggan. *A History of Sicily.* New York: Viking, 1987.

Fischer, Steven Roger. *A History of Writing*. London: Reaktion Books, 2001.

Freeman, Edward Augustus. *The History of Sicily From the Earliest Times*. Oxford: Clarendon Press, 1891–94.

Freeman, Kathleen. *Greek City-States*. New York: Norton, 1950.

Galilei, Galileo. *Dialogues Concerning Two New Sciences*. Edited by Stephen W. Hawking. Philadelphia: Running Press, 2005.

Geikie, Cunningham. *The Holy Land and the Bible*. New York: Cassell, 1887. http://philologos.org/_eb-thlatb/chap31.htm.

Grant, Michael. *The Founders of the Western World: A History of Greece and Rome*. New York: Charles Scribner's Sons, 1991.

Gutas, Dimitri. *Greek Thought, Arabic Culture: The Graeco-Arabic Translation Movement in Baghdad and Early 'Abbāsid Society (2nd–4th/8th–10th centuries)*. New York: Routledge, 1998.

Heath, Thomas L. *Archimedes*. London: Society for Promoting Christian Knowledge, 1920.

————. *A History of Greek Mathematics*. New York: Dover, 1981.

————. *The Works of Archimedes*. New York: Dover, reprint of 1897 Cambridge edition, with 1912 supplement, *The Method of Archimedes*.

Heiberg, J. L. *Archimedis Opera*. 2nd edition. Leipzig: Teubner, 1910–15.

————. "Eine Neue Archimedeshandschrift." *Hermes* 42 (1907): 235–303.

————. *Mathematics and Physical Science in Classical Antiquity*. Translated by D. C. Macgregor. London: Oxford University Press, 1922.

Herodotus. *The Histories*. Translated by Aubrey de Sélincourt. New York: Penguin, 1988.

Herrin, Judith. *Byzantium: The Surprising Life of a Medieval Empire.* Princeton, NJ: Princeton University Press, 2007.

Herrman, Judson. *The New Hyperides in the Archimedes Palimpsest.* 2006. http://www.archimedespalimpsest.org/scholarship_herman1.html.

Hoddeson, Lillian H. "How Did Archimedes Solve King Hiero's Crown Problem?—An Unanswered Question." *Physics Teacher* 10 (1972): 14–19.

Hodos, Tamar. "Intermarriage in the Western Greek Colonies." *Oxford Journal of Archaeology* 18 (1999): 61–78.

Hoffman, Paul. *Archimedes' Revenge: The Joys and Perils of Mathematics.* New York: Norton, 1988.

Hollingdale, S. H. "Archimedes of Syracuse: A Tribute on the 22nd Century of his Death." *Bulletin of the Institute of Mathematics and its Applications* 25 (1989): 217–25.

Jackson, Donald. *The Story of Writing.* New York: Taplinger, 1981.

Jaeger, Mary. "Cicero and Archimedes' Tomb." *Journal of Roman Studies* 92 (2002): 49–61.

James, Peter, and Nick Thorpe. *Ancient Inventions.* New York: Ballantine, 1994

Jean, Georges. *Writing: The Story of Alphabets and Scripts.* New York: Harry N. Abrams, 1992.

Kern, Paul Bentley. *Ancient Siege Warfare.* Bloomington, IN: Indiana University Press, 1999.

Kilgour, Frederick G. *The Evolution of the Book.* New York: Oxford University Press, 1998.

Knorr, Wilbur. "The Geometry of Burning-Mirrors in Antiquity." *Isis* 74 (1983): 53–73.

Kolata, G. "In Archimedes' Puzzle, a New Eureka Moment." *New York Times,* December 14, 2003.

Laird, W. R. "Archimedes among the Humanists." *Isis* 82 (1991): 628–38.

Lange, Lester H. "Did Plutarch Get Archimedes' Wishes Right?" *College Mathematics Journal* 26 (1995): 199–204.

Larsen, Jakob A. O. *"Konig Hieron der Zweite von Syrakus* by Alexander Schenk Graf von Stauffenberg." Book review. *Classical Philology* 31 (1936): 181.

Lawrence, A. W. "Archimedes and the Design of the Euryalus Fort." *Journal of Hellenic Studies* 66 (1946): 99–107.

Lee, Felicia R. "A Layered Look Reveals Ancient Greek Texts." *New York Times*, November 27, 2006. http://www.nytimes.com/2006/11/27/arts/27greek.html.

Levarie, Norma. *The Art and History of Books.* New Castle, DE: Oak Knoll Press, 1995.

Lloyd, G. E. R. *Greek Science After Aristotle.* New York: Norton, 1973.

Lovejoy, Arthur O. *The Great Chain of Being: A Study of the History of an Idea.* Cambridge, MA: Harvard University Press, 1936.

Lowden, John. "Archimedes into Icon: Forging an Image of Byzantium." In *Icon and Word: The Power of Images in Byzantium*, edited by Antony Eastmond and Liz James, 233–60. Aldershot, Hampshire, UK: Ashgate, 2003.

Marsden, E. W. *Greek and Roman Artillery: Historical Development.* Oxford: Clarendon Press, 1969.

Martin, Henri-Jean. *The History and Power of Writing.* Translated by Lydia G. Cochrane. Chicago: University of Chicago Press, 1994.

Martin, Thomas R. *Ancient Greece: From Prehistoric to Hellenistic Times.* New Haven, CT: Yale University Press, 1996.

Merriman, Norman. "Cattle Problem Solved." *New York Times,* January 18, 1930.

Miller, G. A. "Archimedes and Trigonometry." *Science,* New Series 67 (1928): 555.

Netz, Reviel. "Archimedes in Mar Saba: a Preliminary Notice." In *The Sabaite Heritage in the Orthodox Church from the Fifth Century to the Present,* edited by Joseph Patrick, 195–99. Leuven, Belgium: Peeters, 2001.

————. *The Diagrams as Floating Bodies.* http://www.archimedes palimpsest.org/scholarship_netz1.html.

————. "The Origins of Mathematical Physics: New Light on an Old Question." *Physics Today* 53 (2000): 32–37.

————. "Overview: The Importance of the Palimpsest to the Study of Archimedes." http://www.archimedespalimpsest .org/scholarship_netz.html.

————. "Proof, Amazement, and the Unexpected." *Science* 298 (November 1, 2002): 967–68.

————. *The Works of Archimedes.* Volume 1. Cambridge: Cambridge University Press, 2004.

Netz, Reviel, and William Noel. *The Archimedes Codex.* London: Westfield and Nicolson, 2007.

Neugebauer, Otto. "Archimedes and Aristarchus." *Isis* 34 (1942): 4–6.

Noel, William. "Archimedes and Company." *British Academy Review,* issue 10 (2007): 46–48.

————. *The Archimedes Palimpsest Project.* http://www.archi medespalimpsest.org/index.html.

North, J. D. "Seeing Archimedes Through." *Isis* 73 (1982): 271–74.

Oldham, R. D. "Old Archimedes Teases the Moderns." *New York Times,* August 1, 1926.

Olmert, Michael. *The Smithsonian Book on Books*. Washington, DC: Smithsonian Books, 1992.

Osborne, C. "Archimedes on the Dimensions of the Cosmos." *Isis* 74 (1983): 234–42.

Parke, H. W. "A Note on the Topography of Syracuse." *Journal of Hellenic Studies* 64 (1944): 100–102.

Pegg Jr., Ed. "The Loculus of Archimedes, Solved." *American Mathematical Association Online*, November 17, 2003. http://www.maa.org/editorial/mathgames/mathgames_11_17_03.html.

Peterson, Ivars. "Ancient Infinities." *Science News Online* 162 (November 23, 2002). http://www.sciencenews.org/articles/20021123/mathtrek.asp.

———. "Floating Bodies." *Science News Online* 166 (August 21, 2004). http://www.sciencenews.org/articles/20040821/mathtrek.asp.

———. "Turn of the Screw." *Science News Online* 157 (January 22, 2000). http://www.sciencenews.org/articles/20000122/mathtrek.asp.

Plato. *Laws*. Translated by Benjamin Jowett. *The Internet Classics Archive*. http://classics.mit.edu/Plato/laws.5.v.html.

Polybius. *The Histories*. Loeb Classical Library edition, 1922–27. http://penelope.uchicago.edu/Thayer/E/Roman/Texts/Polybius/home.html.

Putnam, George Haven. *Books and their Makers During the Middle Ages: A Study of the Conditions of the Production and Distribution of Literature from the Fall of the Roman Empire to the Close of the Seventeenth Century*. New York: Hillary House, 1962 (reprint of 1897 edition).

Raeder, Hans. "Johan Ludvig Heiberg. 27/11 1854–4/1 1928." *Isis* 11 (1928): 367–74.

Reyhan, Patricia Youngblood. "A Chaotic Palette: Conflict of Laws in Litigation between Original Owners and Good-Faith Purchasers of Stolen Art." *Duke Law Journal* 50 (2001): 955–1043.

Reynolds, L. D., and N. G. Wilson. *Scribes and Scholars: A Guide to the Transmission of Greek and Latin Literature.* 3rd edition. Oxford: Clarendon Press, 1991.

Richmond, Ian A. "*Das antike Syrakus. Eine historischarchäologische Untersuchung* by Knud Fabricius." *Classical Review* 47 (1933): 16–17.

Rickey, Fred. *The Archimedes Manuscript*, 1998. http://www.lix.polytechnique.fr/Labo/Ilan.Vardi/sawit.html.

———. *Re: Archimedes Palimpsest*, October 30, 1998. http://www.lix.polytechnique.fr/Labo/Ilan.Vardi/rickey.html.

Rorres, Chris. *Archimedes.* http://www.math.nyu.edu/~crorres/Archimedes/contents.html.

Rorres, Chris, and Harry G. Harris. "A Formidable War Machine: Construction and Operation of Archimedes' Iron Hand." In *Symposium on Extraordinary Machines and Structures in Antiquity.* Olympia, Greece, 2001. http://www.math.nyu.edu/~crorres/Archimedes/Claw/models.html.

Rose, Paul Lawrence. "Humanist Culture and Renaissance Mathematics: The Italian Libraries of the Quattrocento." *Studies in the Renaissance* 20 (1973): 46–105.

Rouge, Jean. "Greek Colonisation and Women (Tr., Susie Williams)." *Cahiers d'Histoire* 15 (1970): 307–17. (See *Apoikia: The Greek Colonisation Teaching Resource*, University of Liverpool, http://www.apoikia.org.uk.)

Sarton, George. *Six Wings: Men of Science in the Renaissance.* Bloomington, IN: Indiana University Press, 1957.

Schidorsky, Dov. "Libraries in Late Ottoman Palestine between the Orient and the Occident." *Libraries and Culture* 33 (1998): 260–76.

Schreiber, P. "A Note on the Cattle Problem of Archimedes." *Historia Mathematica* 20 (1993): 304–6.

Schulz, Matthias. "The Story of the Archimedes Manuscript." *Der Spiegel Online* (June 22, 2007). http://www.spiegel.de/international/world/0,1518,490219,00.html.

Seid, Timothy W. *Interpreting Ancient Manuscripts*, 2004. http://www.earlham.edu/~seidti/iam/interp_mss.html.

Sevcenko, Ihor. "New Documents on Constantine Tischendorf and the *Codex Sinaiticus.*" *Scriptorium* 18 (1964): 55–80.

Shapiro, A. E. "Archimedes's Measurement of the Sun's Apparent Diameter." *Journal for the History of Astronomy* 6 (1975): 75–83.

Simms, D. L. "Archimedes and the Burning Mirrors of Syracuse." *Technology and Culture* 18 (1977): 1–24.

———. "Archimedes and the Invention of Artillery and Gunpowder. *Technology and Culture* 28 (1987): 67–79.

———. "Archimedes the Engineer." *History of Technology* 17 (1995): 45–113.

———. "Archimedes' Weapons of War and Leonardo." *British Journal for the History of Science* 21 (1988): 195–210.

———. "Galen on Archimedes: Burning Mirror or Burning Pitch?" *Technology and Culture* 32 (1991): 91–96.

———. "The Trail for Archimedes's Tomb." *Journal of the Warburg and Courtauld Institutes* 53 (1990): 281–86.

Sjöqvist, Erik. *Sicily and the Greeks.* Ann Arbor, MI: University of Michigan Press, 1973.

Smith, William. *A History of Greece, From the Earliest Times to the Roman Conquest.* Boston: Swan, Brewer, and Tileston, 1885.

Stein, Sherman. *Archimedes: What Did He Do Besides Cry Eureka?* Washington, DC: Mathematical Association of America, 1999.

Stevenson, Daniel C. *The Internet Classics Archive* (Plutarch, *Marcellus*; *Demetrius*, translated by John Dryden), 2000. http://classics.mit.edu/Plutarch/marcellu.html; http://classics.mit.edu/Plutarch/demetrus.html.

Stille, Alexander. "Resurrecting Alexandria." *New Yorker*, May 8, 2000, 90–99.

Taylor, Isaac. *History of the Transmission of Ancient Books to Modern Times.* New York: Haskell House, 1971.

"Tells Archimedes's Methods of Search." *New York Times*, August 4, 1907.

Thucydides. *The Complete Writings of Thucydides: The Peloponnesian War.* New York: Modern Library, 1951.

Tillotson, Dianne. *Medieval Writing*, 2005. http://medievalwriting.50megs.com/writing.htm.

Tischendorf, Constantin. *Reise in den Orient (Travels in the East).* Translated by W. E. Shuckard. London: Longman, Brown, Green, and Longmans, 1847.

———. *When Were Our Gospels Written?* 2nd edition. New York: American Tract Society, 1867.

Trapp, J. B. "Archimedes's Tomb and the Artists: A Postscript." *Journal of the Warburg and Courtauld Institutes* 53 (1990): 286–88.

Treadgold, Warren T. "The Revival of Byzantine Learning and the Revival of the Byzantine State." *American Historical Review* 84 (1979): 1245–66.

Turner, E. G. *Greek Manuscripts of the Ancient World.* Princeton, NJ: Princeton University Press, 1971.

Twain, Mark. "Archimedes." *Twainian* 12 (November–December 1953): 2–3 (reprint from the *Standard* [Australia], July 27, 1889).

Vardi, Ilan. "Archimedes' Cattle Problem." *American Mathematical Monthly* 105 (1998): 305–19.

———. *The Legacy of Archimedes (287–212 B.C.).* http://www.lix.polytechnique.fr/Labo/Ilan.Vardi/archimedes.html.

von Erhardt, R., and E. von Erhardt. "Archimedes' Sand-Reckoner." *Isis* 34 (1943): 214–15.

Walvoord, Derek J. *Archimedes Palimpsest Character Recognition,* 2004. http://www.archimedespalimpsest.org/pdf/archimedes_c.pdf.

Waterhouse, W. C. "On the Cattle Problem of Archimedes." *Historia Mathematica* 22 (1995): 186–87.

Westlake, H. D. "*Syrakus: zur Topographie und Geschichte einer griechischen Stadt* by Hans-Peter Drögemüller." Book review. *Classical Review,* New Ser. 21 (1971): 97–99.

Whibley, Leonard. *Greek Oligarchies: Their Character and Organisation.* New York: Haskell House, 1971.

Williams, H. C., R. A. German, and C. R. Zarnke. "Solution of the Cattle Problem of Archimedes." *Mathematics of Computation* 19 (1965): 671–74.

Wilson, N. G. *The Archimedes Palimpsest: A Progress Report.* http://www.archimedespalimpsest.org/scholarship_wilson1.html.

———. "Archimedes: The Palimpsest and the Tradition." *Byzantinische Zeitschrift* 92 (1999): 89–101.

———. *Scholars of Byzantium.* London: Duckworth, 1983.

Winter, F. E. "The Chronology of the Euryalos Fortress at Syracuse." *American Journal of Archaeology* 67 (1963): 363–87.

Woods, Heather Rock. "Placed Under X-ray Gaze, Archimedes Manuscript Yields Secrets Lost to Time." Stanford Report, May 19, 2005. http://news-service.stanford.edu/news/2005/may25/archimedes-052505.html.

ACKNOWLEDGMENTS

I am grateful to George Gibson and Jackie Johnson at Walker & Company, who early on pushed me to find the soul of the book; my agent, Sally Brady, who remains an essential and beloved part of my writing life; Professors Toby Huff and Wolfhard Kern, who brought their historical and scientific expertise to the manuscript; my colleagues at the University of Massachusetts Dartmouth, for their continued support and friendship; Harvard College Observatory, for the courtesy appointment which allows me access to Harvard's amazing library resources; and my wife, Sasha, for her patience during all those times when I snuck away to write.

INDEX

A NOTE ON THE AUTHOR

Alan Hirshfeld is a professor of physics at the University of Massachusetts Dartmouth, and an associate of the Harvard College Observatory. He is the author of *The Electric Life of Michael Faraday*, *Parallax: The Race to Measure the Cosmos*, and the *Astronomy Activity and Laboratory Manual*. His essay on Michael Faraday won second prize in the 2005 John Templeton Foundation Power of Purpose essay competition. He lives in Newton, Massachusetts.